Fathoming the Ocean

FATHOMING
the OCEAN

The Discovery and Exploration of the Deep Sea

HELEN M. ROZWADOWSKI

THE BELKNAP PRESS OF
HARVARD UNIVERSITY PRESS

Cambridge, Massachusetts, and London, England, 2005

Library of Congress Cataloging-in-Publication Data

Rozwadowski, Helen M.
Fathoming the ocean : the discovery and exploration
of the deep sea / Helen M. Rozwadowski.
p. cm.
Includes bibliographical references and index.
ISBN 0-674-01691-2 (alk. paper)
1. Underwater exploration.   2. Ocean bottom—Research.
3. Oceanography—History.   I. Title.
GC65.R695 2005
551.46—dc22   2004057083

Designed by Gwen Nefsky Frankfeldt

For my parents

Contents

Foreword by Sylvia A. Earle

BY NATURE, humans are curious, always trying to figure out what's around the next bend in the road, over the next hill, beyond the next star—or under the surface of the dominant feature of Earth, the sea. Inevitably, the ocean will be explored from the most hostile frozen seas to the warmest, most secluded tropical lagoons and ultimately, to the greatest, darkest depths, seven miles down, but as this engaging volume by Helen Rozwadowski clearly shows, access below a few feet underwater will always rely on harnessing technologies to overcome our limitations as terrestrial, air-breathing mammals.

Imagine what it would be like to fly in a plane slowly, thousands of feet in the sky above Boston, London, Sydney, or Tokyo, a thick blanket of clouds obscuring the view below. What could be discovered about such places if you had to rely on time-honored oceanographic techniques—lowering nets or open-ended metal boxes from high in the sky, then dragging such devices across the surface of whatever is below or dropping baited hooks fastened to long lines to see if some unwary creature could be enticed to bite and be taken from its realm into ours for careful examination? For many decades, methods such as these were used to try to piece together knowledge of the ocean by observers sailing aboard naval ships, private yachts, and, eventually, dedicated ocean research vessels.

In eloquent prose, Rozwadowski brings alive the personalities and daunting problems of pioneering ocean explorers who faced the challenges of trying to probe the blue part of the planet without the benefit of modern acoustics, electronics, and sophisticated materials to answer basic questions: How deep is the ocean? Can life survive where no light penetrates? What is the nature of ocean currents? Much of the information that forms the basis of modern oceanography has been literally dragged, hauled, hooked, and pulled, one data point at a time, from the depths of the ocean, usually from the deck of a rolling ship, by people using an amazing array of ingenious inventions.

Scientists today often refer to the famous global voyage of the British vessel *HMS Challenger* from 1872 to 1876 as the beginning of modern oceanography, but Rozwadowski provides well-documented examples of many less well celebrated precursors to that historic voyage, and tells hair-raising stories of dangers that were overcome to extract small but vital bits of knowledge needed to determine safe sailing routes or find appropriate places to lay transoceanic cables in the 1800s. It hasn't been—and still is not—easy to get answers. Even now, less than 5 percent of the ocean below a hundred feet or so has been seen, let alone explored. The wonder is not how little is known about the ocean, but rather how much, considering the difficulties until recently of getting around on the surface, and how limited access is today, even to the average depth of the ocean, two and a half miles down. A dozen astronauts have walked on the moon and hundreds have felt the weightlessness of space travel, but only two people have ventured and safely returned from the deepest crack in the ocean, the Challenger Deep at the bottom of the Mariana Trench.

More than two centuries ago, Benjamin Franklin pieced together information provided by sea captains to produce the first sketch of the Gulf Stream, a prominent feature in the Atlantic Ocean that previously had escaped definition. During the century that followed, the time that is the primary focus of this book, it is safe to say that more was learned about the ocean than during all preceding human history. The same can be said for the twentieth century. Today, satellites provide daily updates on the ever changing configuration of the Gulf Stream and numerous other aspects of the ocean's surface, from temperature and the nature and extent of plankton blooms to wave height and even the migration patterns of certain large sea animals. Sensors aboard satellites measure ripples that can be analyzed to determine wind speed and direction. From high in the sky it is even possible to determine the peaks, valleys, and plains of the sea floor thousands of feet underwater based on slight variations in the height of the sea surface that faintly mirror the configuration of the terrain below.

Although some ships still deploy nets, trawls, and dredges to determine the nature of the ocean below, the limitations of such blind sampling are giving way to methods that take observers directly into the three-dimensional depths personally, using various small manned submersibles, or take them vicariously, with numerous remotely operated vehicles that provide images and streams of data, sometimes from land-based centers many miles from where the undersea vehicles are deployed. Since the 1960s, people have been able to live underwater in special undersea dwellings, sometimes staying for weeks at a time, using the ocean itself as a laboratory for in situ observations and experiments. Military submarines with many people aboard are capable of staying submerged for weeks, traveling the oceans of the world, even under polar ice. Oil and gas industries worldwide scan with acoustic devices that probe and map the structure of the deep ocean floor. Plans are under way to develop a worldwide network of monitoring stations, a "Global Ocean Observing System," that will eventually make possible improved weather forecasting as well as greatly extending and expanding knowledge about the ever changing nature of the ocean.

In the twentieth century, discoveries were made about the ocean that would have astounded explorers of only a few decades earlier. The existence of 40,000 miles of mountains that run like giant backbones down the major ocean basins at last was recognized, as was an explanation for the movement of continents and oceans through plate tectonics. Direct observations and photographs of many living creatures in the deepest seas forever put to rest the idea that life there is not possible. The discovery of hydrothermal vents spouting hot water not only rich in minerals but also teeming with microbes revolutionized scientific thinking about the nature of the planet's geological and biological processes. Recent insight concerning the astonishing diversity and abundance of microbes in the sea, from the newly recognized kingdom Archeae to bacteria and viruses, is provoking new ways of looking at and thinking about the origin and future of life.

A vital, overarching discovery of the twentieth century came as a consequence of all the individual shreds of knowledge at last assembled in ways that clarified the significance of the ocean to humankind. We now recognize that the ocean is the engine that drives the way the world works, shaping climate and weather, governing planetary chemistry, generating most of the oxygen in the atmosphere, absorbing much of the carbon dioxide, providing living space for most of life on Earth, in terms of both genetic diversity and sheer mass. The ocean, after all, harbors 97 percent of Earth's water, the single nonnegotiable thing life requires.

Some say the great era of exploration is now over, that for new "ocean frontiers" we must look beyond our own atmosphere, perhaps to Mars, where there once was an ocean, or to Europa, one of Jupiter's moons, where liquid salt water may flow and may host something alive under a thick mantle of ice. Yet the example provided by the heroic characters portrayed in this book should inspire another view: the past is prelude, with every new discovery opening dozens of new doors to the next levels of understanding.

On the basis of all that had been discovered before, by the middle of the twentieth century Rachel Carson was inspired to write in *The Sea Around Us* : "Eventually man . . . found his way back to the sea . . . And yet he has returned to his mother sea only on her own terms. He cannot control or change the ocean as, in his brief tenancy on earth, he has subdued and plundered the continents." As a new century dawns, we now know that, in fact, it *is* possible to significantly alter the nature of the sea through what we have put into it—hundreds of millions of tons of noxious wastes—and what we have taken out—hundreds of millions of tons of wildlife. According to recent, well-documented studies, 90 percent of the big fish—tuna, swordfish, marlin, sharks, cod, grouper, snapper, and many others—are gone, and most of the other ocean species sought after as food are in a state of serious decline. Coral reefs, sea grass meadows, mangrove forests, and other coastal systems that once seemed infinitely resilient have declined abruptly, and globally, in recent decades. Around the world, pollution has given rise to more than a hundred coastal "dead zones."

Knowledge that humankind does have the capacity to alter the nature of the sea may be the most important discovery made so far about the ocean. But the greatest discovery may await, of learning how to live within our planetary means. Thanks to generations of curious, daring, intrepid explorers of the past, we may know enough, soon enough, to chart safe passage for ourselves far into the future.

Fathoming the Ocean

chapter one

Fathoming the Fathomless

We do not associate the idea of antiquity with the ocean, nor wonder how it looked a thousand years ago, as we do of the land, for it
was equally wild and unfathomable always. The Indians left no
traces on its surface, but it is the same to the civilized man and the
savage. The aspect of the shore only has changed. The ocean is a
wilderness reaching round the globe, wilder than a Bengal jungle,
and fuller of monsters, washing the very wharves of our cities and
the gardens of our sea-side residences.

—Henry David Thoreau, *Cape Cod,* 1864

THE BEACH was deserted, the sea empty of sails. The retreating storm clouds still stirred the water, but without the ferocity that wrecked the brig *St. John* and cast the bodies of would-be Irish immigrants on the Cape Cod beach in October 1849. Since local townspeople and grieving relatives had retrieved beached corpses—although not all of the 145 lost souls—only driftwood collectors and seaweed harvesters had ventured onto the beach. Even the inshore fishermen remained home. Yet Concord native Henry David Thoreau walked along the edge of the ocean, gazing out at the wind and waves.

Beaches, such as the one at Newport, Rhode Island, were already a popular destination for city dwellers seeking the healthful effects of sea breezes and bathing, but not such desolate stretches of coast as the shores of Cape Cod. Thoreau spent several weeks, over three trips, tramping from Eastham, a town just north of the elbow of Cape Cod, to Provincetown on its tip. He deliberately sought out this unfashionable, sand-covered stretch, peopled by families whose habits and livelihoods were foreign to inland citizens. Like many educated, middle-class people of his day, Thoreau was interested in natural history and carefully examined the shellfish, birds, and beached whalebones he discovered. He lamented the absence of scientific information on whales and seals in the books he consulted, and he described with relish the shore whaling he witnessed. Most of all, he "went to see the Ocean," which in different moods he called "savage," "unwearied," "illimitable," and "fabulous."

Thoreau's rambles and musings on Cape Cod demonstrate an awareness, new in the mid-nineteenth century, that the ocean was an unmarked and unstudied realm, especially in contrast to land, which had long yielded to civilization and to scientific scrutiny. Yet the years from 1840 to 1880 witnessed a dramatic increase in awareness of the open ocean as a workplace, a leisure area, a stage for adventure, and a natural environment. The Anglo-American world enjoyed a deepening acquaintance with the seashore, with the maritime culture of the high seas, and also with the ocean's depths. Because the deep sea could only be known indirectly, through fishing, whaling, or attempts to dip sampling devices beneath the waves, it is hardly surprising that the same decades also saw the emergence of scientific interest in the depths.

Before the nineteenth century, the deep sea made hardly any impression on most people, even citizens of maritime nations. Mariners rarely strayed from proven trading routes or familiar fishing grounds. Even ocean-going explorers were more land than ocean oriented; they used the sea merely as a highway to get to the next landfall. The mid-nineteenth-century Anglo-American world witnessed a surge of interest in the high seas on the part of virtually all classes. Starting in the previous century, Europeans discovered the beach, whose waters and air they sought for health reasons. Appreciation of distant natural landscapes from the upper-class Grand Tour extended to the seascape. As beach holidays became a regular part of the lives of even the middle and lower classes, more people than ever before began setting sail across the blue water. At a time when more people were working at sea than at any time previously, emigration and ocean travel for pleasure also exposed landlubbers to the experience of seagoing. When voyagers returned to dry land they remained conscious of the ocean, which recurred in their everyday lives in sermons, stories, and pictures.

As seagoing became an important part of British and American culture in the nineteenth century, a strong tradition of writing about voyaging emerged. Although writing had long played a central role in maritime life and work, particularly for a handful of educated explorers and voyagers, middle-class willingness to go to sea translated into an efflorescence of maritime writing, both published and unpublished, by working sailors and whalers, professional writers, ships' officers, and scientists who set sail. The new genre of maritime novels appeared in the same decades that men of science turned their attention seaward. Like middle-class writers, scientists went to sea with preconceptions inspired by explorers' narratives. From midcentury, they also carried with them ideas about seagoing instilled by novelists who described voyaging as, among other things, a route to self-knowledge. Ocean scientists chose to report on their activities not just in papers in scientific journals but also in popular voyage narratives. This preference provides a glimpse of their motivations for setting sail.

Along with their literary and laboring counterparts, middle-class scientists sought the experience of voyaging, with all the opportunity, danger, heroism, and self-transformation attendant to the new nineteenth-century encounter with the ocean. Their choice of genre also reflected their belief that ocean study deserved large-scale government support commensurate with that afforded the nationalistic explorations that formed the subject of many popular voyage narratives.

The first scientific attention paid the deep sea coincided with interest in similar sorts of areas. The atmosphere, the Arctic, the jungle, the underground, and the deep sea all offered an almost unimaginably vast three-dimensional scale. Getting to these places, and surviving there, presented daunting technological challenges. These environments imposed strict physical limitations on human investigators, whose scientific understanding of them was necessarily mediated through complex technologies. The nineteenth century saw a general increase in efforts by scientists to comprehend and control large spaces, through meteorology, scientific ballooning, the study of terrestrial magnetism, Arctic exploration, and mountaineering. Ocean scientists looked for inspiration to these allied fields as they struggled to understand the far reaches of the globe that became their purview. Like other contemporary field sciences, early ocean science blended the promise of tangible economic benefit with the political potency that derived from mapping and discovering.

The story of the oceans at midcentury is not merely the penciling in of a previously blank chart. It is instead the tale of the expanding human imagining of what the "deep sea" might be. Until the first decades of the century, the location of the ocean's bottom—if it existed at all—was anyone's guess. The 1823 *Encyclopedia Britannica* entry for "Sea" stated simply, "Through want of instruments, the sea beyond a certain depth has been found unfathomable." Although navigators needed only to rule out shallowness, several enterprising groups of people set sail at midcentury to discover and explore the ocean's depths.

BEFORE THE nineteenth century, the deepest parts of the ocean existed only as unfathomable barriers between places or as watery highways bounded by waysides as "blank and . . . untraveled and as much out of the way of the haunts of civilized man as are the solitudes of the wilderness that lie broad off from the emigrants' trail to Oregon."[1] The point of setting sail was always to get back to land as soon as possible. Sailors respected and feared the ocean. Landlubbers rarely thought about it, but when they did, their attention had usually been attracted by the news of a shipwreck, mutiny, or other disaster. Europe under the influence of merchantilism treated the ocean as nonpossessable space wherein states could, nevertheless, extend state power over particular trade routes. By the late eighteenth century, Europeans understood the deep sea as a great void that was empty and featureless, the antithesis of civilization.[2]

During the nineteenth century, the blue water became a destination rather than a byway or barrier. It became a workplace on an entirely new scale, and a place amenable to control, or at least comprehension, by technology and science. Whalers began to hunt sperm whales in the deep sea. Explorers pushed far into the Arctic regions. Increasing passenger travel, emigration, and shipping boosted sailing packet, clipper ship, and steamer traffic across the Atlantic.[3] In response to increased shipping, British and American hydrographic offices conducted active programs to chart domestic, colonial, and foreign ports. They also experimented with deep-sea sounding, especially after submarine telegraph cable promoters called for conquering the Atlantic's greatest depths.

By the mid-nineteenth century, the ocean had become a resonant reference for people who did not work at sea or live along the coast. The widespread popular encounter with the sea was grounded firmly in the powerful British and American maritime economies. For Britain, mastery of the sea went hand in hand with its island geography. With the growth of a colonial empire, it came to seem inevitable that the British navy ought to rule the waves. English self-definition as an ocean-oriented nation existed before the nineteenth century. It set the stage for ready acceptance of an idea that arose starting in the 1840s: that the deep ocean was an important place and a natural site for the exercise of British military, technological, and scientific power. This impulse seemed amply confirmed by the success of submarine telegraphy in the 1860s.[4]

The simultaneous function of the Atlantic as a bridge and a moat shaped American government, commerce, and culture. New England's economy had been based on whaling and trading long before the nineteenth century. Maritime commerce financed westward expansion and taught business and management skills that were applied to land activities. The maritime economy promoted interdependence with Europe, but distance encouraged Americans to develop their own customs. The persistence of nautical references in the sermons of Puritan ministers suggests how deeply the crossing experience influenced those who endured it. The ocean barrier strengthened migrants' original commitment to America and compelled new Americans to look to the future rather that dwelling on the past.

Crossing the ocean gave immigrant Americans a shared historical past, which meant that even in an age of land expansion maritime interests maintained a currency throughout the country.[5]

Interest in the deep ocean began at the sea's edge. Before the last quarter of the eighteenth century, understanding of the ocean's depths derived mostly from the imagination. Stories from ancient literature and the Bible shaped the image of the sea more than accounts of travel or exploration. The seashore was known only as a scene of disaster, peopled by cannibals, mutineers, and shipwreck victims. Robinson Crusoe, for example, rarely ventured onto the beach. Between 1750 and 1840, that changed, when "the desire for the shore . . . swelled and spread."[6]

An early sign of admiration of the sea was the mid-eighteenth-century Grand Tour visit to Holland to view the Dutch seascape made famous in paintings. The conscientious tourist of the late eighteenth century brought along on his adventures a watch, compass, astrolabe, and field glasses, as well as an artist if he could afford the expense. An integral part of the journey entailed sketching and otherwise recording the traveler's emotions and observations, with the ultimate aim of producing a personal travel account. Those who visited the seashore sought the sense of the sublime that arose from actual experience; they felt the "agreeable amazement" generated by the extreme calm or exquisite violence of the water. Romantics celebrated the sea as a transcendent realm beyond progress, civilization, and development. For Romantic artists, the seashore became the ideal place for personal reflection and self-knowledge because of the correspondence between marine and psychological depths. Grand Tourists, though, remained ignorant of the perils of the sea; to them, storms were merely pictorial elements. Early upper-class admirers of the sea were clearly insulated from reality.[7]

Not only did the coast serve as a setting for self-reflection, but it also became a place where observers could grasp the new sense of time proposed by geologists. As the history of humankind became disconnected from that of the planet, the idea of a very ancient earth, indifferent to human presence, achieved sublimity. Observers began to perceive the coast, rocks, and cliffs as products of age-old wear. The shore simultaneously held images of past, present, and future. Cliff views offered three-dimensional spectacles that satisfied observers and readers who were becoming accustomed to new three-dimensional representation. People came to the coasts to "browse in the archives of the earth." Reflecting on endless waves and the indistinct shoreline, they could envision the eternity of the world.[8]

The appeal of the sea was not limited to aesthetics. Sea-bathing came into

Children, some dressed in sailor suits, and passengers returning from a sail, at Brighton Beach, 1884. (Hulton Archives / Getty Images.)

vogue in the mid-eighteenth century, arising from the aristocratic practice of retreating to the countryside in the face of disease or threats to the nobility's political or social power. Merely breathing sea air was considered healthful, especially compared to the fetid and disease-ridden air of cities in the summer. Bathers sought cures for melancholy, bad temper, and anxiety through strictly regulated hydrotherapy. The fashion for the beach rested on the paradox that the sea was a refuge and source of hope because it inspired fear. People enjoyed the sea yet endured the terror it engendered in order to overcome their infirmities. Under the guise of medical therapy, genteel bathers experienced a new world of bodily sensations.[9]

The model of the Brighton bathing holiday owed much to inland spas such as Bath. In Britain, both were initially the preserve of aristocrats who followed members of the royal family there. Genteel seaside holiday practice, initially limited to the very upper classes, expanded in concentric circles to include celebrities, a wider part of the gentry, and then industrial and merchant middle classes. For these groups, a stay at a fashionable beach resort was more more easily attainable than the high society life of English country houses. Places like Brighton allowed less wealthy gentry to verify their social status and aspiring middle classes to find suitable marriage partners. When the latter groups arrived at the shore, they added their own rhythms and customs to their imitations of the royal model. Bathing, bookshops, reading rooms, boutiques, visits, walks, and boat outings allowed young people to escape from close supervision. Through the 1830s, the social life of the great English resorts such as Cowes and Brighton remained focused on the activities of the aristocracy, but starting around 1840 working-class visitors appeared in resort towns. In that decade, railroads unleashed crowds on the beaches and made the ocean accessible to everyone. By the 1840s, then, the English had discovered the seashore. American discovery, which soon followed, likewise depended on the confluence of railroads, therapeutic bathing, the educational benefits of natural-history collecting, and desirable society.[10]

In the next several decades landlubbers' attention shifted from the beach to the ocean's depths far beyond. Ocean travel and emigration gave thousands of people direct experience with the open sea for the first time in history. Before the start of the nineteenth century, few people who did not work at sea traveled on long voyages. Only emigrants and colonial officials had reason to go, and few of them made more than one or two trips in their lives. Not only was going to sea dangerous, but it involved an uncomfortably close encounter with an international maritime culture of life at sea that polite society preferred to ignore. From the perspective of shore, those people who worked at sea, especially common seamen, represented the dregs of society.[11] During the nineteenth century, though, shipping and whaling increased dramatically, so that greater numbers of people, including middle-class men, began to work at sea. In addition, emigrants and

A steamer leaving Liverpool, with the tender pulling away as friends and family wave good-bye. Trans-Atlantic travel became increasingly fast, luxurious, and predictable in the third quarter of the nineteenth century. (*Illustrated London News,* Feb. 19, 1881, 188–189.)

travelers, including scientists, artists, and writers, took to sea in greater numbers than ever before.

By midcentury, fleets of scheduled packet ships filled the sea lanes. Regularly scheduled sailing ships began service in 1818 between Britain and the United States. American lines dominated trans-Atlantic passenger and mail service until British companies took over the market with steam vessels. The Cunard line began service in the 1840s, almost twenty years after the steamship *Savannah* first crossed the Atlantic. The *Savannah* used its machinery only in sight of audiences on the shore. The first transoceanic journey made entirely under steam was the twenty-six-day crossing of the *Royal William* in 1831. At that time, the better sailing packets traversed the Atlantic faster, but within a decade Cunard's *Britannia* crossed in thirteen days. In 1850 the American Collins line *Atlantic* made the voyage to New York in only ten days and sixteen hours. For the next decade, American lines tried to establish a niche in the market. The Collins line succeeded in attracting 50 percent more passengers than the Cunard line. The Collins line's success, based on a "reputation for fashion, dash, and a broad hint of recklessness," presaged its downfall. A series of wrecks and lost ships sent passengers flocking back to Cunard, with its more sedate image. Still, speed and luxury continued to compete with safety. By the third quarter of the century, the fastest ships could cross the Atlantic in just over a week.[12]

At first, packet passengers were aristocrats, but in response to emigrants' growing demand cargo space was converted into quarters for "steerage" passengers by the installation of rough bunks. Unable to afford passage on steamers, most of the seven and a half million people who crossed the Atlantic between 1800 and 1870 traveled on sailing ships, living in communal dormitories built into cargo spaces five or six feet high. Until midcentury reforms, they endured weeks, or even months, of the squalor and stench associated with extreme overcrowding, minimal sanitation, lack of ventilation, and poor food. The Inman line then turned its attention to the emigrant demand and became the first to offer affordable fares in steamships. By the 1870s, even emigrants traveled in steamers. Second-class accommodation, introduced in the 1850s, broadened opportunities for affordable and respectable ocean travel.[13]

In spite of the sheer number of emigrants, midcentury steamship companies focused their advertising, amenities, and publicity unwaveringly on first-class passengers. With regularly scheduled service, people with the means to travel could do so conveniently. Increasingly, they could also do so comfortably and fashionably. A passenger of the White Star line remarked, "The *Republic* is a floating palace, with the style and comfort of a Swiss hotel." Sea travel ceased to be a sometimes miserable, often boring, and yet life-defining adventure of uncertain duration. Suddenly it became reasonably safe, relatively predictable in length, and even comfortable by the standards of the day. Efforts to cater to passengers' comfort and entertainment produced the grand "traveling palaces" that ushered the elite back and forth across the Atlantic in style, and with increasing speed.[14]

Large crowds attended launchings of new steamships, whose arrivals and departures were newsworthy. Newspapers reported crossings by celebrities with enthusiasm, as when the Swedish singer Jenny Lind, sponsored by P. T. Barnum, voyaged to the United States on the steamer *Atlantic*. Writers such as Jules Verne and Charles Dickens reported their impressions of the crossing to popular audiences. Dickens's experience so horrified him that he chose a sailing packet for the return voyage. Verne sailed on *Great Eastern*, a technological wonder five times larger than any ship afloat at the time. Delighted with his experience, he compared the promenade deck to

Jules Verne compared a stroll on the deck of the *Great Eastern* to a Sunday afternoon walk on the fashionable Parisian boulevard the Champs Elysée or in London's Hyde Park. This leviathan, five times larger than any other ship afloat, set out on its maiden voyage in 1860 but never achieved the promise envisioned by its creators, except as the vessel that finally laid the trans-Atlantic telegraph cable in 1866. (*Harper's Weekly*, Aug. 7, 1859, p. 222; courtesy of Archives & Special Collections at the Thomas J. Dodd Research Center, University of Connecticut Libraries.)

the Champs Elysées or Hyde Park on a fine Sunday afternoon in May. By the 1880s, ocean liners provided such luxurious settings that wealthy elites began to make ocean-liner crossings part of the yearly round of high society.[15]

Atlantic travel remained uncertain, though, and missing and wrecked ships became grist for shoreside rumor mills. Periodicals easily sold to a public eager to be thrilled and frightened by pathetic and gruesome stories. Lost ships starkly reminded potential travelers and anxious relatives of passengers of the danger and unpredictability of the ocean. Seasickness was a more mundane but constant reminder that the ocean was not a natural human environment. The fancy saloons, private ladies' parlors, gentlemen's smoking rooms, and grand dining halls, which provided relief from the monotony of passengers' tiny private cabins, also distracted them from the fact that they were at sea. Ships offered more and more amenities, including electric bells, running water, bathtubs, oil lamps, and steam heat. By the last quarter of the century, publicity brochures boasted that travelers could pay for "the privilege . . . of seeing nothing at all that has to do with a ship, not even the sea."[16]

When passengers did not insulate themselves from the sea, ocean crossings provided them with an education in life and work at sea. From the start of scheduled packet service, women as well as men sailed as passengers on long voyages. On sailing vessels, male passengers, especially those in steerage, relieved the boredom of long, empty hours at sea by helping with the work of the ship. Women usually remained belowdecks, although they often participated in rat hunts. Steerage passengers were not far separated from the crew, either in type of accommodation or proximity to the daily work and discipline of the ship. Thus passage across the Atlantic by sail gave emigrants a glimpse of what life at sea was like. On some ships, first-class passengers felt free to engage in the class tourism of visiting the crew's living quarters.[17] Sea travel, whether for pleasure or for emigration, introduced millions of landsmen and landswomen to the otherwise unfamiliar, even alien, maritime culture.

Immigrants to the United States almost never boarded ocean liners again after they arrived in America. Instead, they took their experiences with them as they forged new lives, often in the West. But emigrants and returning passengers did not escape reminders of the sea. They formed the audience for news about exploring expeditions and the newly popular sea stories. Most spectacularly, the attempts in the 1850s and 1860s to lay a transatlantic submarine telegraph cable riveted landlubbers' attention. Evocations and images of the sea abounded at midcentury. Hobbies, clothing, collections, reading materials, even restaurant menus reflected a deepening acquaintance with the ocean.

The submarine cable venture illustrates both the growing level of awareness of the deep ocean and one of the most powerful commercial and political reasons that study of the deep ocean accelerated when it did. Before the first attempt to lay a cable across the Atlantic, the longest underwater cable had been 110 miles and the greatest depth in which cable was laid 300 fathoms (1 fathom is 6 feet). The Atlantic cable would have to be 2,000 miles long and placed in water up to three miles deep, a feat many believed impossible.[18] After an unsuccessful attempt in 1857, in which the cable broke about halfway across, another attempt in 1858 succeeded in laying a cable from Ireland to Newfoundland only to have it fail a month later. Because

The USS *Niagara* (left) and the HMS *Agamemnon* (right) in a gale during the trans-Atlantic cable-laying voyage of 1858. Each ship carried half the cable. A midocean splice and the successful landing of the two ends, in Newfoundland and Ireland, were greeted with wild enthusiasm by the press and observers waiting for news in both countries. The 1858 cable failed after about a month, though, because of electrical problems. (*Harper's Weekly*, Aug. 14, 1858, p. 521; courtesy of Archives & Special Collections at the Thomas J. Dodd Research Center, University of Connecticut Libraries.)

of financial difficulties, recalcitrant electrical problems, and the American Civil War, the next attempt was not made until 1865. That summer, the cable broke again. Finally in 1866, a cable was successfully laid, by the leviathan *Great Eastern*, from Valencia, Ireland, to St. John's, Newfoundland. As soon as the crew finished laying that cable, they returned to pick up the 1865 cable and completed it as well. Deep-sea surveys preceded each set of attempts to lay the Atlantic telegraph. Those expeditions, sponsored by the American and British navies, represented some of the first attempts to sound the deep ocean and resulted in a dramatic increase in knowledge about the deep-sea floor. Most important, they established the need, in the eyes of the American and British governments and scientific institutions, for further investigation of the deep ocean.

Events surrounding the attempts to lay submarine cables received widespread public attention from the press. The successes of 1858 and 1866 elicited effusive congratulations; the failures brought forth criticism from naysayers. The idea of a wire linking America and Britain resonated in the public imagination. More than simply a business tool, submarine cables came to seem strategically invaluable, an absolute insurance of peace forever. Cable surveys drew particular attention to the deep-ocean floor, whose profile and constitution came into focus before an amazed audience as hydrographers probed the depths and translated their measurements into pictures. Among the new findings that impressed the public, especially the scientific community, was the discovery of a multitude of unfamiliar creatures attached to the Mediterranean cable raised for repair in 1860 from about 1,000 fathoms. The previously academic debate over the azoic theory proposed by the British zoologist Edward Forbes, which posited that no life existed below 300 fathoms, spilled over into a society that embraced

This allegorical scene shows Neptune in the foreground in front of the cable held by Britain, represented by the lion, and the United States, represented by the eagle. In the center at the top is a portrait of Cyrus Field. The Atlantic submarine telegraph cable, often touted as the eighth wonder of the world, was hailed as a communications breakthrough that would ensure peace between nations. (Published by Kimmel & Forester, 1866; Library of Congress, Prints & Photographs Division, LC-USZC-2388, Digital ID: cph 3g02388.)

The first trans-Atlantic yacht race in 1866. Three American schooners raced for a prize of $60,000. (Published by Currier & Ives, 1867?; Library of Congress, Prints & Photographs Division, PGA—C&I—Great Ocean Yacht Race.)

natural history in various forms as entertainment. As a result of the cable voyages, the public, including men of science, absorbed much knowledge about the deep-sea environment.[19]

This knowledge included the striking discovery that the recently unfathomable depths could be profitably commodified. In the first flush of excitement over the successful laying of the 1858 cable, Cyrus Field sold twenty miles of the leftover cable to Tiffany & Company Jewelers. Tiffany cut some of it up into four-inch pieces, attached certificates of authenticity, and sold them for fifty cents as souvenirs. The company transformed other sections into umbrella handles, canes, and watch fobs. Cyrus Field, entrepreneur of the Atlantic telegraph, had a watch fob made with grains of deep-sea sediment set in it. An enterprising chemist in New York advertised a perfume created in Field's honor, "distilled from ocean spray and fragrent flowers."[20]

The same year in which a working trans-Atlantic cable was finally laid, another high-seas drama entertained spectators on the shore. In December of 1866, the first trans-Atlantic yacht race pitted three American schooners against each other, with a purse of $60,000. A decade and a half earlier, the schooner *America* had sailed across the Atlantic to challenge members of the Royal Yacht Squadron. When the English yachtsmen saw how fast *America* sailed while cruising, most declined to race. The ones who did lost, much to the glee of the Americans and the consternation of British yachtsmen. Eighteen years after the initial race, the surviving owners of the cup awarded the winning yachtsmen designated it a trophy for the international competition still known as the America's Cup. For the general public, yacht racing provided an entertaining reason to look seaward. Although not the most popular Victorian spectator sport, yacht races nevertheless provided occasions for betting and were widely reported in the press.[21] For the yachting elite, time spent at sea allowed them to embrace their own, carefully sanitized version of maritime culture.

The vogue of dressing children in sailor suits was probably inspired by the model of royal yachtsmen. From the founding of the first yacht clubs early in the century, yachtsmen adopted fancy nautical garb. In 1854, Queen Victoria's son Prince Arthur was reported to have appeared during "The Week," the major summer regatta at Cowes, dressed in his first sailor suit, just as his father, Prince Albert, had before him in 1846. From midcentury on, English parents began dressing their little boys in sailor suits, a fashion that lasted for over a century. They chose a style that more closely resembled clothing worn by eighteenth-century sailors than the uniforms worn by contemporary ones. The look spread to other countries, so that by 1885 the sailor suit in its various versions was one of the most widespread styles for both boys and girls. Women, too, adopted nautically inspired costumes, initially in yachting circles and later outside them.[22]

During the nineteenth century, then the ocean entered the minds, homes, dreams, and conversations of ordinary people. The high seas became a place of travel and adventure, while the ocean's depths were revealed as a fascinating natural environment. So familiar did the ocean's water become that advertisers felt free to invoke "deep-sea" as a sales tactic. The ichthyologist George Brown Goode told his Fish Commission boss Spencer Baird, "In Boston I investigated the 'deep-sea flounder' on the bill of fare at Youngs." But by this time scientific knowledge of the sea encompassed deep-water fauna, so Goode could report scornfully, "It's nothing in the world but *Chaenopsetta ocellares* and by no means so palatable as our friend *G. cynoglossus.*"[23]

⚓ THE MOST comprehensive introduction to the deep ocean resided in the pages of nineteenth-century books about ocean voyages. Literature about the sea appealed to a growing market of readers with a taste for it.

Maritime writers found ready audiences among populations becoming ever more familiar with ships, wrecks, seafaring language, the names of small islands in the Pacific, and even the shapes and habits of marine creatures. As literacy rates climbed in the nineteenth century, stories about life at sea became staples of adults' and children's reading. As maritime literature increased in popularity, it served to introduce the nonseafaring public to the world of ships and voyages. Maritime literature figured prominently in American and British cultural experience and helped both countries construct images of themselves as maritime nations.[24]

Maritime literature, both published and unpublished, both written at sea and penned retrospectively, relied on the format of a voyage narrative: the chronological story of a voyage, sometimes constructed like a journal with separate daily entries. The widespread popularity of voyaging literature, both fiction and nonfiction, rendered the genre of voyage narrative an obvious choice for all seagoers who wrote about their experiences.

Long before the nineteenth century, writing was an integral part of maritime work. Most, though not all, ships on long-distance voyages kept logbooks, which were important for navigational as well as legal reasons. Log entries might, for instance, be invoked as evidence in the case of a mutiny. Efforts at record keeping varied by industry. Whalers reported daily positions only casually, but they noted carefully where and when they saw or took whales. Sealers, by contrast, often did not keep logbooks or destroyed them after the voyage, to keep the location of profitable islands secret. Fisherman, who tended to return to the same, familiar grounds, did not usually keep logbooks. By the mid-eighteenth century, trainees on British naval vessels were required to keep personal logbooks as a qualification for advancement. Explorers and surveyors, often naval officers, kept very detailed logbooks to support their claims of discovery and priority.

Record keeping achieved greater prominence on oceangoing vessels during the nineteenth century. This reflected the contemporary revolution in organization of government and industry, whereby large bureaucracies extended control over far-flung enterprises. Keeping written accounts of work done and supplies used became part of the job description for naval petty officers. On a U.S. naval ship in 1826, each section—including the gunner's, boatswain's, sailmaker's, and carpenter's, in addition to the master's—ordered ruled books, foolscap paper, quills, ink powder, inkstands, lead pencils, slates, and slate pencils.[25] A ditty on the cover of a logbook from the same period enjoined its keepers:

> May they whose Lot this Log to keep
> Be worthy of the Task complete

And never leave a sentence out
Which should occur the voyage about—
But always true, and faithfull be
In every Port; as well at Sea,
And every land that should appear,
Mark down their bearings and distances here.
Likewise the winds as they prevail
And whether snow, rain, or Hail'
Rocks and shoals you must include
Their views, depths, and Lattitude.[26]

Rising literacy rates in both the United States and Britain, countries with large and fast-growing maritime sectors, meant that more working sailors and passengers could read and write. The expansion of hydrographic surveying early in the century brought a new, writing-intensive kind of work to sea, involving extensive composition, calculation, and sketching. When marine zoologists set sail on naval vessels, they met hydrographers and other officers who, like themselves, combined strenuous outdoor activity with desk work to record, analyze, and communicate their observations.

After logbooks, the most common form of maritime writing was journals. Sailors, usually officers but also sometimes common seamen, frequently kept journals or wrote reminiscences after retirement. Many passengers on ocean voyages also kept diaries to stave off boredom. So did whalers, whose hard work alternated with idle periods. For explorers, both sailors and scientists, journal keeping was an important part of the work. On the United States North Pacific Exploring Expedition, all officers and scientists were required to keep journals, which became part of the official record. These accounts could also be transformed into lucrative publications.

Writing at sea satisfied personal as well as professional needs. First-time sailors kept journals to record the novelty of their experience for themselves and their families and friends. Old salts wrote to occupy their time. Whaling voyages in the nineteenth century could last four years or longer. Not all diarists spelled out their reasons for writing as clearly as the British naturalist Thomas Henry Huxley, when he confided to his fiancée, "I had always written [my journal] as an amusement in weary hours and for my own eye, solely. It is an unconnected confusion of many thoughts, a laying bare of my private mind."[27] Life at sea alternated between exhausting physical work or danger and tedium. Endless hours standing watch, staring at a constant horizon that could, nevertheless, change unpredictably, inspired reflection.

The idea that voyages provided ordinary sailors with an occasion for Romantic reflection was new in the nineteenth century. Inland middle-class boys who yearned to go to sea were inspired not only by the promise of financial reward and exhilarating physical challenge, but also by an impulse for self-improvement that characterized their class. Many young men who shipped as whalers embraced Victorian values of self-control, temperance, punctuality, sexual continence, and thrift.[28] For these sailors, as for seagoing scientists, writing offered an appropriate medium through which to filter their experiences as ocean voyagers.

Models for writing about ocean voyages included Grand Tour travel accounts and narratives of exploring expeditions. To prepare for the Grand Tour, young men read existing travel accounts, which provided a touchstone against which they could measure their own experiences. Keeping a diary of their own travels was indispensable for conscientious tourists. Although most did not publish their accounts, many travelers wrote journals thinking that they might someday do so. Just as the experience of the Grand Tour lent itself to writing about it, ocean travelers in the nineteenth century often felt compelled to write about their adventures, perhaps partly because of the association of visits to Holland with Romantic admiration for the sea. Early travel accounts, along with Dutch seascape paintings, explained to readers, and to future travelers, not only how to experience the sublimity of the seashore but also how express the encounter.[29]

When educated travelers began to embark on ocean voyages in the nineteenth century, they prepared for the journey by reading published accounts. For sea travel, this meant consulting narratives written by explorers who had discovered new lands and peoples. This popular, literary form shared with Grand Tour accounts the didactic quality of coaching voyagers about what to expect in foreign lands. In addition, these narratives invited voyagers to participate in the act of discovery by chronicling their own personal experiences of little-known areas of the globe. Shipboard journal writers often described their encounters with natives, or with local fauna and flora, as an extension of past Western contact with that particular geographic place. When Charles Darwin described the sound of the language spoken by the natives of Tierra del Fuego, he said that "Captain Cook has compared it to a man clearing his throat."[30]

Besides travelers' and explorers' accounts, nineteenth-century voyagers also had access to the new literary form of the maritime novel. Novels themselves were new in the late eighteenth century, but the maritime world became a literary setting only when the ocean entered British and American consciousness. The motif of the voyage offered a natural framework for stories, and tales of voyaging could explore both the geographical and the psychological unknown. For readers, sea stories provided rare glimpses into the otherwise unfamiliar world aboard naval, merchant, and whaling vessels. The retired British sea captain Frederick Marryat published books in the 1810s and 1820s that have been described as the first maritime novels. America's first two successful professional writers, Washington Irving and James Fenimore Cooper, wrote well-received sea stories. Thus in the second quarter of the nineteenth century, the sea became an inexhaustible source of inspiration and promise for writers and readers.[31]

Herman Melville and Richard Henry Dana were among the first generation of writers to base stories upon personal experience of the sea. Before then, although ex-sailors wrote about ocean voyages, educated landlubbers would not have voluntarily faced the danger of sea travel. Midcentury maritime novels were often written by men in their late teens or early twenties with a significant amount of education, usually at the college level. They came from the middle or upper classes and thought of themselves as gentlemen, especially in contrast to the seamen who became their workmates. They went to sea having read voyaging accounts that encouraged them to think of their own seagoing as part of an ongoing cultural and literary tradition.

Maritime literature of the nineteenth century was a nostalgic product of the brief period of time when the sea held enough romance and mystery to fire the imagination but less threat than in previous centuries. A curious asynchronicity exists between certain themes in sea literature and historical reality. By the 1870s, writers and folklorists were chronicling the heroic "last days of sail" while the clipper ships tried to maintain a place for sailing vessels in a world that had largely converted to steam power. As late as the 1880s and 1890s, when steam had largely supplanted sail even on the long hauls, myth making was alive and well. The image of the "glorious" last days of sail was created largely by retired seaman-writers as well as by authors of maritime novels and other voyage narratives who imbued seagoing with psychological and spiritual value.[32]

Literature played an important role in redefining seagoing, depicting it not as an unwelcome necessity but as a desirable experience for middle-class young men. Maritime novelists brought to public attention the deplorable working and living conditions of common sailors. Richard Henry Dana wrote with a reformer's pen, intending to educate the American public about the brutality and ugliness of maritime work. Ironically, although he hoped to warn young men away from going to sea through his realistic portrayal of shipboard horrors, his book instead inspired many to ship out.

Naval and merchant marine reforms began around midcentury in both the United States and Britain, largely as a result of an outcry from a public newly educated about the realities of life at sea. Better living and working conditions in turn made life at sea more appealing for middle-class men who harbored expectations that sea voyages offered economic opportunities along with adventure.[33]

This first generation of seamen-writers defined the act of going to sea as a growth experience, an opportunity for young men to test their mettle. By the end of the nineteenth century, in the stories of Joseph Conrad and others, the ocean had been irrevocably transformed into a testing ground of a person's physical and psychological strength. Cape Horn, a place that did not achieve immortality until it was tamed, best exemplifies this transformation. Before the nineteenth century, few literary references to Cape Horn exist, and none by authors or poets who experienced it. Its heyday as a ship route, from 1845 to 1855, came only after several important achievements: ships large enough to make the voyage relatively safely, proper food to keep crews healthy for the long passage, and accurate charts. Dana's famous novel, *Two Years before the Mast,* immortalized the passage around the Horn. Getting ready to round the cape—in the form of refurbishing warm clothing and overhauling the vessel's sails and rigging as well as listening to the stories of old-timers describing the terrors to come—set the scene so that the narrator, and also the readers, understood how much more formidable were the waters there than any previously encountered on the voyage.[34] In Herman Melville's "Introductory to Cape Horn" in *White-Jacket,* the narrator reviewed the history of navigation around the Horn. He acknowledged accounts based on these voyages as "fine reading of a boisterous March night." But, he continued, "if you want the best idea of Cape Horn, get my friend Dana's unmatched *Two Years Before the Mast.* But you can read, and so you must have read it. His chapters describing Cape Horn must have been written with an icicle."[35]

Since Dana and other midcentury writers, Cape Horn has come to symbolize the unbridled power and danger of the ocean. Almost as soon as it became accessible to clipper ships, though, further technological changes began to eliminate it as an important sea route. By 1869, with the advent of the Suez Canal, the transcontinental railroad, and steamships that could navigate the Strait of Magellan, traffic around Cape Horn all but stopped, ceasing completely with the opening of the Panama Canal.[36] Yet the Horn retained its literary value, continuing to symbolize man's encounter with the sea as a personal test much more than it represented the historical act of traversing a particular ocean route.

Like writers, scientists began to feel safe and comfortable at sea at mid-century, and they likewise embraced the literary culture of voyaging. This is true not only of the first generation of self-styled ocean scientists, but equally of their predecessors: explorers, hydrographers, submarine tele-graph engineers, marine naturalists, and yachtsmen and yachtswomen. All of these groups relied on voyage narratives and other literary sources for cues about how to interpret their own experiences at sea. The extent of their debt to the literary tradition of oceangoing became apparent in the extent to which they referred to, and borrowed from, other mariners' accounts when constructing their own. Along with other voyagers, scientists participated in the construction of the ocean as a stage for personal growth and mastery of nature.[37]

Novelists, explorers, common sailors, and scientists all utilized the natural structure of voyages to frame their stories. More surprisingly, so did yachtsmen-naturalists who dredged during summer holidays to collect marine invertebrates and hydrographers who conducted deep-sea cable surveys. Published and unpublished writings, both those produced at sea and those composed long afterward, reveal that nineteenth-century voyagers shared a common conception of the appropriate literary forms for conveying the experience of ocean travel.

At the end of a career "before the mast" that included sailing with the U.S. Exploring Expedition under Charles Wilkes, the sailor Charles Erskine reflected, "I have battled the ocean's storms for twenty years and am aware that no language can give reality to the story of my experience." Erskine's emphasis on experience of the sea, and the artistic difficulties of conveying that experience, indicate that he counted himself among the new nineteenth-century population of respectable, upwardly mobile seagoers. He noted, in explaining his decision to go to sea, that his family had been financially well off until his father left them. His imitation of Dana's famous novel in the title of his book, *Twenty Years before the Mast,* suggests a conscious effort to participate in a cultural, literary encounter with the sea. Like Dana and other contemporary seagoing authors, Erskine recorded his experience on the Wilkes expedition as a voyage narrative, whose daily entries offered readers vicarious participation in the voyage. The realism implied by this literary form was belied by the fact that Erskine learned to read and write during the voyage and could not have based his book on a shipboard diary written entirely during the expedition.[38] Erskine's choice of journal entries emphasizes the extent to which they seemed a natural and appropriate format for relating the experience of voyaging.

The literary product of the first trans-Atlantic deep-sea sounding survey provides a striking example of how pervasive the voyage narrative format was for maritime writing. From 1851 to 1852, Lieutenant Samuel P. Lee made observations of wind and currents for Matthew Fontaine Maury's navigational studies conducted from the USS *Dolphin*. Lee's duties included taking a series of deep-sea temperature measurements and soundings, the latter made to confirm or disprove the existence of vigias (reported shoals that endangered navigation in otherwise deep water). The sounding results proved valuable not just for hydrography but for science and for submarine telegraphy. For any of these purposes, a terse report accompanied by a bottom profile would have effectively communicated Lee's findings. Instead he devoted over one-third of his 300-page report to a day-to-day narrative of the voyage.[39]

The report began with an introduction that described the vessel and its equipment, named the officers, reproduced the instructions, and reviewed the track covered by the *Dolphin* during the cruise. These features characterized all exploring expedition accounts and hydrographic reports, regardless of how long or short they were. The rest of the voyage narrative consisted of entries that described day-by-day events: the functioning of equipment, sounding methods used, mishaps, sail handling, food rationing, crew assemblies, and miscellaneous observations of flying fish and seabirds. Possibly Lee employed this journal format to give future deep-sea hydrographers the benefit of detailed information about his deep-sea sounding gear and methods, although, if so, he also included much extraneous information. Perhaps he, Maury, and even Congress, considered the *Dolphin* voyage a national achievement worthy of this type of permanent record, a modest version of an exploring expedition narrative. In the end Congress validated the importance of Lee's report, and by implication, perhaps, the appropriateness of the genre he chose, by ordering 2,000 extra copies to be printed.[40]

Other deep-sea sailors joined Lee in employing the voyage narrative format for reporting their adventures and accomplishments. Although deep-sea sounding did not continue, after Lee's report, to inspire book-length narratives, submarine cable-laying ventures were particularly suited for story telling because of their built-in drama. As every mile of cable was paid out, the stakes rose should the cable break and not be recovered. Each of the Atlantic cable-laying voyages in 1857, 1858, 1865, and 1866 was chronicled in day-to-day accounts in newspapers and periodicals. Some of these appeared afterward as books, such as John Mullaly's *Laying of the Cable*. Participants also kept personal accounts of the voyages, including Charles Bright, who later published his version as *The Story of the Atlantic Cable*.[41]

Anne Brassey writing in her journal aboard her family's yacht *Sunbeam* during their eleven-month voyage in 1876–77. (From Anne Brassey, *Around the World in the Yacht "Sunbeam"* [New York: Henry Holt, 1878], 423.)

Cruising yachtsmen and yachtswomen sometimes published accounts of their adventures. For example, Edward Frederick Knight, a pioneer of yacht cruising who wrote many books about sailing, published accounts of his voyages in his twenty-eight-foot yawl *Falcon*. Anne Brassey chronicled her family's wanderings around the world aboard the yacht *Sunbeam* in book form. More and more tales by yachtsmen appeared around the turn of the century, typified by the well-known story about crossing the Atlantic by Joshua Slocum. Even earlier in the century, though, naturalist-dredgers who used yachts for cruising and collecting wrote narrative accounts of their voyages. Often these had quite technical titles, such as John Gwyn Jeffreys's "On the Marine Testacea of the Piedmontese Coast." Like Jeffreys, other yachtsmen-dredgers published their voyage accounts not as books but in scientific journals, often those of local natural-history societies. Thus Robert MacAndrew wrote "An Account of Some Zoological Researches Made in British Seas during the Last Summer" for the proceedings of Liverpool's Literary and Philosophical Society.[42] Even in these modest forums, and in spite of the primary intention to communicate information about marine zoology, these yachtsmen presented their scientific results in the context of stories about their cruises.

Voyagers adopted the genre of the voyage narrative not only because it offered a natural structure for their stories, but also because they relied on information contained in similar accounts written by navigators, explorers, and whalers. Nineteenth-century scientists shared a common language with other educated writers and readers. Novelists and men of science not only read each other's work but also made use of each other's ideas and metaphors. Maritime novelists relied on published and unpublished whalers' stories to lend verisimilitude to their work. Melville is a particularly good example because he based much of his work on narratives published by whalers. In 1841, while he worked on a Pacific whaler, he acquired from a seaman on a passing ship a copy of that sailor's father's story of a huge whale who sank the ship that attacked it. Even the name of Melville's great white whale, Moby-Dick, was derived from fact; an immense bull sperm whale christened Mocha Dick terrorized whaleboats from 1810 until it was finally killed in 1859, with nineteen old harpoons embedded in his body. Several books about sperm whales written by surgeons on British whaling vessels provided background for Melville's famous whaling story; one of them also contributed to J. M. W. Turner's 1840s whaling paintings.[43]

The first generation of ocean scientists, too, relied on information from whalers, who unlike most mariners remained in blue water for long periods instead of crossing it quickly. Whalers alerted navigators to unusual or new features in the sea because they strayed from established shipping lanes. Matthew Fontaine Maury, for example, actively sought observations from whalers and other mariners to produce information about the depths. He collected logbooks, whale sightings, and meteorological and hydrographic measurements to create sailing directions, charts, and his 1855 book, *The Physical Geography of the Sea*.[44] The British microscopist George Wallich, who sailed with the cable-surveying voyage of HMS *Bulldog*, bolstered his argument that animals could live at great depths with a story of Captain William Scoresby's about a sperm whale dragging a boat down to such a depth that the wood from the boat sank like a stone for a year. He concluded that if whales could survive the pressure at such depths, so could other animals.[45] Writers such as Wallich bridged the social gulf between the maritime world and the polite society of men of science by incorporating sea stories into scientific works.

As scientists learned more about the sea, novelists read and utilized ocean scientists' work in a similar way. Jules Verne's imaginative undersea voyage reflected the discovery of the depths especially well. Inspiration for *20,000 Leagues under the Sea* derived from a visit to an aquarium and the experience of crossing the Atlantic on the immense, self-contained world of the famous steamship *Great Eastern.* While aboard, Verne interviewed crew members who had just the previous year participated in the laying of the Atlantic telegraph cable. He also met Cyrus Field, principal promoter of the cable venture. Verne himself was an ardent yachtsman and wrote most of this book while cruising on his yacht *Saint-Michel.* He crafted his novel with a copy of Maury's *Physical Geography of the Sea* beside him. The route that the fictional *Nautilus* took around the world reflects the order in which Maury's book treats geographic regions of the ocean. Whole passages in Verne's story mirror Maury's text. Verne's reliance on Maury shows that his fanciful adventures were embedded in state-of-the-art geography and science.[46]

Extensive borrowing from literary sources particularly characterized the voyage narratives that ocean scientists wrote. The first generation of ocean scientists to sail on naval vessels adopted the genre of exploring expedition narratives. Although these scientists set sail to investigate the sea, most of them devoted their reminiscences to port stops, just as other explorers had before them. As voyagers keeping journals, or those writing retrospectively, approached a port either literally or discursively, they consulted other accounts to learn about the island or area they would visit. Often they copied passages into their journals or summarized other accounts to provide readers with background information on the geography, political history, and native people of the region. When HMS *Challenger* headed for Bermuda early in its 1872–1876 expedition to study the world's oceans, several of the scientists sought information about the island. In his published account, Charles Wyville Thomson included a long history of when and by whom Bermuda was discovered and settled, followed by an account of its geography and geology. Against this backdrop he presented his own experiences, including the remarkable sighting of a sand glacier that had engulfed a house, leaving only a chimney showing. Places with less colorful histories received less literary attention. Thomson's description of Halifax, Nova Scotia, merited only four pages while Bermuda had inspired fifty-seven. Although most often background information derived from published geographies and explorers' narratives, voyagers also relied on their own ship's logbook as well as shipmates' journals. Lieutenant Pelham Aldrich devoted a short section to "Remarks on Bermuda," including statistics of trade exports and imports, yearly temperature and climate information, and the island's latitude and longitude, all of which he copied from Navigating-Lieutenant Thomas Henry Tizard's journal.[47]

Challenger scientists and officers used explorers' voyage accounts to help them make sense of their experiences at sea. Many of them had, like

Drawing of the HMS *Challenger* made fast to St. Paul's Rocks, six hundred miles northeast of Natal, Brazil, by the expedition artist J. J. Wild. (Published in Charles Wyville Thomson and John Murray, ed., *Report of the Scientific Results of the Exploring Voyage of the HMS "Challenger," 1873–76: Narrative of the Cruise,* vol. 1 [London: HMSO, 1885], 201; courtesy of Dr. David C. Bossard.)

the naturalist Henry Moseley, read Darwin's narrative of the *Beagle* and Daniel Defoe's story of Robinson Crusoe when they were young. For Moseley these books fanned his desire to travel across oceans. Before he sailed with *Challenger* he had already accompanied an eclipse expedition. Lieutenant W. J. J. Spry likewise carried the memory of Robinson Crusoe's adventures with him to sea. At the island of Juan Fernandez he reflected on the experience of solitude, remembering that Alexander Selkirk had spent four years alone there before his rescue in 1709, after quarreling with his captain and requesting to be put ashore on the uninhabited island. Back in England, Selkirk's story had inspired Defoe to write his famous tale, *Robinson Crusoe.* The chief scientist on the *Challenger,* Charles Wyville Thomson, wrote home to his son about the Juan Fernandez visit, knowing it would capture the boy's imagination. Another stop, at the island group of Tristan da Cunha in the South Atlantic, prompted Thomson to write for his published voyage narrative a history of the discovery and settlement of the islands. Lieutenant Andrew F. Balfour likewise included a long description of the islands in his personal journal. Perhaps this port call was especially memorable for expedition members because they rescued two Germans who had lived on Inaccessible Island, part of the Tristan da Cunha group, for nearly two years.[48]

Not only did *Challenger* officers and scientists interpret their adventures in the light of well-known voyage narratives, but they actively tried to make their own experiences fulfill the expectations these accounts raised. Thus as *Challenger* approached St. Paul's Rocks, the crew read accounts of previous visits to the rocky islets 600 miles from Natal, Brazil. Thomson studied the voyage narratives of James Clark Ross and Charles Darwin. Lieutenant Aldrich, an enthusiastic sportsman, noted from reading the same books that the rocks were an excellent spot for shark fishing. From Darwin he gleaned advice about the best bait and hooks to use, which he prepared in advance. For him and others, the stop at St. Paul's Rocks met their expectations of what exploring expeditions should be like. First the crew performed the remarkable technical feat of making fast to the rocks. They then spent several enjoyable days shooting birds, fishing for sharks, and collecting specimens.[49] Like Aldrich, Victorian seagoers were well schooled in the art of voyaging. Their behavior at sea was directed by the narratives they read to prepare for their travels. In turn, they incorporated their maritime experience into their own voyage narratives, helping thereby to shape the expectations and experiences of the next generation of sailors.

Ocean scientists gravitated to the traditional genre of explorers partly to bring their work to an interested public who could experience the sea through their narratives. By rhetorically placing themselves in the role of great explorers and discovers who charted and named the far corners of the world, scientists transformed the deep sea into the next frontier. The ocean joined other frontiers that attracted the attention of nineteenth-century explorers and scientists who, having discovered and claimed most of the earth's land masses, turned to the poles, the atmosphere, and the oceans.

As SCIENTISTS struggled to find ways to conceptualize the ocean, they sought analogies to other field sciences, especially astronomy and geology. The deep-sea sounding pioneer Matthew Fontaine Maury compared the work he promoted to other, more familiar kinds of investigation. He compared the behavior of the Gulf Stream to the orbit of planets. In particular, he related study of the sea to astronomy: "There were in the depths untold wonders and inexplicable mysteries. Therefore the contemplative mariner . . . experience[d] sentiments akin to those which fill the mind of the devout astronomer when, in the stillness of the night, he looks out upon the stars, and wonders." Because Maury communicated his discoveries both to scientists and to mariners, he made an effort to explain his investigations in terms that all his audiences would understand. He promoted his work by claiming that its practitioners belonged "alongside of the navigator, the geologist, and the meteorologist, with a host of other good fellows." To emphasize the practicability of his program to study winds and currents, Maury again relied on the analogy to astronomy, asking, "Could it be more difficult to sound out the sea than to gauge the blue ether and fathom the vaults of the sky?" Maury's reference to astronomy served multiple purposes. It asserted the ocean's depths as equally deserving of reflection and scrutiny as the heavens. It also provided a model for studying a vast, seemingly unreachable place by insisting that, if human ingenuity could reach the "vaults of the sky," it could certainly fathom great depths. Third, it presented ocean science as akin to one of the most respected scientific pursuits of the day. By equating ocean science with astronomy, Maury attempted to provide for it the cultural position of an intellectual, scientific pursuit that nevertheless retained a practical orientation.[50]

Other ocean scientists also remarked on the resemblance of their work to astronomy. To describe the seabed to readers of his *British Conchology*, John Gwyn Jeffreys explained, "if it could be viewed through an aquatic telescope, [its outline] would be seen to be irregular, and nearly as much diversified as the surface of the earth." After four years of sailing and studying the oceans, Henry Moseley reversed the analogy to express his new understanding of the earth and the sky. "After a voyage all over the world, there is nothing which is so much impressed on my mind as the smallness of the earth's surface . . . We live in the depths of the atmosphere as deep-sea animals live in the depths of the sea." William B. Carpenter, one of the two men most directly responsible for the *Challenger* expedition, cited eclipse expeditions as a precedent for government support of scientific voyages.[51]

Carpenter also drew comparisons between the sea and the polar regions, an association that many proponents of ocean science made. Deep-sea investigators especially presented their work as a descendant of the acclaimed early nineteenth-century Arctic explorations that searched for the North Pole and the Northwest Passage. They saw a close relationship between Arctic exploration and natural history that made it seem evident that their work followed in this tradition. Edward Forbes noted that Arctic explorers were more attuned to natural history than other explorers: "Whether it is that the paucity of objects for observation induces attention to such as are seen, or that the cold air sharpens men's wits, voyagers to the Arctic seas, not being professed naturalists, have paid much more attention to the animals which inhabit them, and described those they have met with much more intelligibly [sic] than travelers in warmer and more favoured climates."

While ocean scientists looked to established sciences such as astronomy or famous undertakings such as Arctic expeditions, other field scientists drew analogies between vast, inaccessible sites. Ballooning allowed adventurers to cross "oceans" of air. The scientific balloonist James Glaisher turned seaward to explain why ballooning remained valuable even with the looming prospect of heavier-than-air flight: "What would be said of a sailor who, a hundred years ago, abandoned his sailing vessel because he had a faint notion of steamboats?"[52]

The comparisons that ocean scientists, and other contemporary field scientists, drew to make sense of their undertakings were more than mere analogies. Many eighteenth- and nineteenth-century attempts to comprehend the natural world seemed to promise tangible economic benefits. Astronomy was the science behind improvements in navigation in the eighteenth century, most notable of which was the development of instruments and methods for determining longitude at sea. Geology grew out of prospecting and mining work in the eighteenth century and maintained an interdependence with it. Then geology went on to gain government support successfully because of the conviction on the part of many scientists and politicians that geological surveys, which discovered and mapped resources, would more than adequately repay the initial investment.[53] The same was true of the newer field sciences. Terrestrial magnetism proved valuable to mariners trying to navigate by compass, especially after the advent of iron-hulled ships. Tidal investigations promoted trade and overseas expansion by providing mariners and navies with accurate information about the shores of maritime nations and their colonial possessions. Meteorology promised the means for safer ocean travel as well as the possibility of predicting ocean harvests. Studying the sea promised to shed light on the problems of declining fisheries, especially whaling, while hydrography appeared poised to guarantee safer navigation.[54]

Although potential commercial benefits promoted many, perhaps most, field sciences, motives for systematic investigation of vast portions of the globe were at least as political and cultural as economic. Only after most of the world's coastlines had been at least roughly charted and most of the islands claimed in the name of one of a handful of Western nations did explorers and scientists turn their attention to the Arctic, the atmosphere, and the ocean's depths. The synchronous emergence of polar exploration, glaciology and mountaineering, meteorology, and ocean science is hardly coincidental. These scientific undertakings represented to mid-nineteenth-century explorers and scientists tangible and symbolic ways to extend their control of territory and their knowledge of the globe. At that time, British and American navies bolstered national policies to establish spheres of economic and political interest around the globe. Astronomy and botany participated in this endeavor by directing ships to the far corners of the world to set up permanent and temporary observatories and to find and relocate valuable plants.[55]

Investigation of the deep sea accrued power to the nations whose ships and citizens set sail. At a time when it seemed to many observers that all commercially important continents and islands had been discovered and claimed, scientific discoveries or technological feats demonstrated cultural superiority equally as well as "showing the flag." Midcentury deep-ocean hydrography promised to open up the depths to transoceanic telegraph cables. Politicians quickly recognized that the nation which could lay, repair, and sabotage cables at will would rule more than the waves.[56] Sounding great depths extended the pursuit of precision and mathematical description that characterized general surveying work. Even before telegraphy became a motive, inscribing a numerical depth on the charts offered a similar kind of personal achievement and national prestige as discovering a new island, inlet, or anchorage.

The functional depth of the ocean changed dramatically over the course of the nineteenth century. Navigators had no use for sounding lines greater than 100 fathoms, while explorers before the mid-eighteenth century carried only 200-fathom lines. Seamen disagreed about whether an object thrown overboard, including a sounding lead, would sink to the bottom or come to rest in some depth at which the water's density matched its own. Were this the case, soundings that reached bottom would be physically impossible. Even after hydrographers believed they could accurately measure the deepest parts of the ocean, this idea persisted. Efforts to lay the first Atlantic submarine telegraph cable in 1857 and 1858 were preceded by popular speculation that the cable might not lie on the bottom but rather might float at some intermediate depth. As late as the *Challenger* voyage of the mid-1870s, the bluejackets, or common sailors, sent a delegation to ask the scientists whether the body of their shipmate Bill, just buried at sea, would "find his level" or sink to the bottom. Although the sailors' ignorance amused the scientist who recorded this query, the level of scientific knowledge about the conditions at great depths hardly justified condescension. Later during the voyage, the scientists found it necessary lower a live rabbit down on a 500-fathom line in order to observe the condition of the body after exposure to great pressure.[57]

The sea was brought into focus as an object of scientific inquiry by two groups of people, the hydrographers who charted its depths and the naturalists who studied its inhabitants. Both began close to shore, probing blindly with sounding devices and dredges into ever deeper water. Until the 1860s, they maintained divergent definitions of what, in fact, constituted deep water. There were efforts to extend the range of human divers, particularly in association with the construction of bridges and tunnels in industrial centers, but the maximum depth for effective work by humans under water remained in the 75- to 120-foot range until the turn of the twentieth century, and this technology was not much employed by scientists.[58]

The first systematic efforts to sound the open sea date from the mid-nineteenth century. A handful of eighteenth-century explorers attempted deep-sea measurement, as did Arctic explorers of the early nineteenth century, most particularly Sir John Ross and his nephew Sir James Ross, who reported one sounding of 1,000 fathoms. Until midcentury, despite such individual and piecemeal efforts, Maury lamented, "the bottom of what the sailors call 'blue water' was as unknown to us as is the interior of any of the planets of our system."[59]

Beginning in the 1850s, deep-water measurements formed a small part of the regular missions of hydrographic vessels that traversed blue water on route to their assignments.[60] It was also the sole charge for a handful of government-funded surveys for various telegraph cables. Results in the early years regularly revealed depths in the 1,000- to 2,000-fathom range.[61] But a few significantly deeper measurements were also recorded. The deepest sounding from Captain Samuel Barron of the *John Adams,* in the Atlantic between Virginia and the Madeiras in 1851, was 5,500 fathoms. Captain Henry Denham sounded 7,706 fathoms in 1852. In 1842, depths up to about five miles had been considered possible, on the assumption that the ocean's basins are proportional to those on land. As a result of these few early measurements, men of science concluded that the greatest depth of the sea was at least eight to nine miles.[62]

Soon such extremely deep measurements came to be considered dubious. By the 1870s, hydrographers judged as improbable the existence of depths greater than 5,000 fathoms. A British hydrographic publication summarized: "as a general rule the depths are under 3000 fathoms; depths in excess of 3000 fathoms have with one or two exceptions been alone attained north of the Equator, and it is worthy of note that the exceptional deep holes of about 4000 fathoms have hitherto been found unexpectedly and within a short distance of land." Accordingly, George E. Belknap, the hydrographer on the US Steamer *Tuscarora* in 1874, considered a 3,054-fathom sounding "deep" while describing 2,500 fathoms as "moderate depths."[63]

Compared to hydrographers, the vast majority of naturalists who studied marine fauna and flora experienced the sea at more modest depths, though no less challenging for their equipment and experience. Although references exist to Italians dredging in shallow water with an ordinary oyster dredge in 1750, and to the Danish naturalist Otto F. Müller rigging a naturalist's dredge for collecting in 30 fathoms at about the same time, a continuous tradition of natural-history dredging dates only from the 1830s. Depths of 20 fathoms seem to have been notable at that time, but the next decade saw some hauls around 100 fathoms.[64]

Whereas hydrographers tended to push the limits of their equipment and expertise to sound increasingly deeper water, many marine naturalists were content, until the 1860s, to use dredging as an extension of beachcombing, a way to reach into the sea just past the low-tide line, perhaps to tens of fathoms. Even when dredgers wanted to work farther out at sea in the late 1840s, their desires were easily deflected, as when the naturalist Albany Hancock explained a change of plans in 1847: "Mr Flower has not been out dredging since you left, the weather has been so uncertain. And I am afraid he will not be able to go to the Dogger [a distant and relatively deep fishing bank, by naturalists' standards] this year . . . I have recommended him therefore to take a small boat and try the coast nearer at hand . . . He will be able to dredge 3 or 4 times for the cost of one Dogger's bank expedition, and on the whole will probably be just as productive."[65] There was no pressing need for naturalists to reach farther down, because most of them could get hauls rich enough during a few summer rowboat excursions to keep themselves busy all winter.

Later, a few naturalist-dredgers became interested in collecting mostly, even exclusively, from greater depths than those sampled before. During the North Pacific Exploring Expedition in the mid-1850s, most of William Stimpson's dredging took place in 10 to 30 fathoms. Near the end of the voyage, he could reach 30 to 50 fathoms. In 1852, one British naturalist-dredger wrote to a colleague who requested a particular shallow-water species, apologizing for "my only dredging having been in deep water."[66]

Not until the 1860s did naturalists' definition of "deep" water change substantially. By 1861, John Gwyn Jeffreys, who spearheaded British dredging efforts in that decade, spoke of dredging in "deep water" at 86 fathoms. Jeffreys spent part of every summer collecting around the northern fringes of the British Isles, working regularly in the 80-fathom range and, in 1867, reaching 170 fathoms. By 1869, Jeffreys wrote to *Nature* objecting to use of the term "deep sea" for only 10 fathoms, which had been the naturalists' definition in the 1840s. He declared, "For such depths as those explored at the present day no term short of 'abyssal' was appropriate," and suggested labeling 50 fathoms "deep sea." By 1877, he had further stretched the definition, identifying "abyssal" as anything from 100 to 1,000 fathoms, reserving the term "benthic" for greater depths.[67]

The desire to explore great depths grew stronger as scientists found new and ever more unusual fauna in deep water. In 1864, the American zoologist and son of the famous naturalist Louis Agassiz, Alexander Agassiz, wrote excitedly of the results of a series of deep-sea dredgings accomplished by the U.S. Coast Survey: "The Fauna living to a depth of 500 fathoms = 3000 feet—*10* atmospheres! is *Wonderfully Rich* in EVERY- THING Echinoderms, Corals, Ophiurans, Starfishes, Annelids, Crustacea Mollusca, etc." During that decade naturalists teamed up with hydrographers to reach with their dredges depths previously accessible only to sounding devices. U.S. Coast Survey ships dredged depths up to 850 fathoms during the special voyages deployed to study the Gulf Stream in the late 1860s. During the summer of 1868, HMS *Lightning* dredged at 650 fathoms. The following year that accomplishment was dwarfed by HMS *Porcupine's* successful haul at 2,435 fathoms. By 1870, naturalists referred to depths of 80 to 100 fathoms as "trifling" and considered the 300- to 800-fathom range "what we now must call moderate depths."[68]

chapter two

The Undiscovered Country

Beneath the waves, there are many dominions yet to be visited, and kingdoms to be discovered; and he who venturously brings up from the abyss enough of their inhabitants to display the physiognomy of the country, will taste that cup of delight, the sweetness of whose draught those only who have made a discovery know.

—Edward Forbes, *The Natural History of the European Seas,* 1859

T HE RENOWNED navigators and explorers who sailed during the Great Age of Discovery of the fifteenth and sixteenth centuries mapped passages between oceans to newly discovered lands and islands. But not even these explorers dallied on the high seas. The desire to travel safely and quickly between places drove the great navigational advances of the eighteenth century, including the crucial development of chronometers for determining longitude at sea. The first mariners to sail off the tracks discovered by early explorers and long retraced sea routes were adventurers: whalers searching for offshore species, sealers seeking unknown islands where they could work without competition, and Arctic explorers sent out in part to bolster the sealing and whaling industries. Sperm whalers in particular followed their prey far from land and watched in amazement as harpooned whales pulled hundreds of fathoms of rope down as they sounded. The stories these men told when they returned to shore conveyed to both scientists and the public a new awareness of the vastness of the ocean's depths.[1]

To those intent on opening up the deep ocean, eighteenth-century explorations provided an obvious and compelling model. Captain James Cook established a tradition of incorporating science into voyages of exploration and discovery. Participating in scientific exploration demonstrated military prowess, commercial muscle, and cultural superiority. The North Pacific Exploring Expedition of 1853–1856, for example, staked a U.S. claim to the depths by focusing its scientific attention on marine organisms and deep-sea sounding experiments. Although a continuous tradition of scientific investigation of the depths did not emerge, fully formed, out of geographic exploration, the tradition of exploration did make critical contributions to the nineteenth-century definition of the ocean as unclaimed territory over which nations exerted cultural influence and political control.[2]

Societies and nations have not extended control over the oceans in the same way as over land. The Indian Ocean before European influence, for instance, was viewed as a realm set apart from society, mysterious and untamable because of the monsoons that limited trade to certain seasons. Micronesians, by contrast, viewed the ocean as territory that provided connections between places as well as food and other resources. By the early nineteenth century, the open sea was understood by Europeans as a wild place and a great void outside society, yet one that provided links between places. By midcentury, geophysical scientists had transformed the vast emptiness of the sea's surface into an ordered and bounded grid defined by the tides, magnetic fields, and temperatures. Next, men of science pondered the ocean's third dimension. Deep water, the sea floor, and marine creatures became not only objects of scientific interest but also the means investigators used to assert national claims to this new territory. Submarine telegraphy would soon transform scientific control achieved through cultural imperialism into the more concrete uses of the sea floor made by transoceanic cables.[3]

WHEN SCIENTISTS took to sea in significant numbers, many competing technologies and methods for learning about the depths emerged: Sounding, dredging, microscopy, temperature measurements, chemical analysis, and laboratory experiments all contributed to the developing nineteenth-century understanding of the deep sea. Long before that, information about the ocean derived from navigators, explorers, fishermen, and whalers. Contrary to the image of ships setting sail over a "trackless waste," Matthew Fontaine Maury asserted that "the winds and the currents are now coming to be so well understood, that the navigator, like the backwoodsman in the wilderness, is enabled literally 'to blaze his way' across the ocean."[4] Sperm whalers, though, often strayed from the invisible highways followed by conservative mariners. Simply by virtue of pursuing their prey beyond known pathways, whalers became the vanguard of ocean science.

As Deputy Postmaster General for the American colonies, Benjamin Franklin turned to whalers to find out why British ships sailing a northern route across the Atlantic from England took longer to cross than vessels sailing a southern route. He brought the problem to his cousin Timothy Folger, a Nantucket ship captain familiar with the whaling industry. Folger immediately understood that the merchant and packet ships fought against

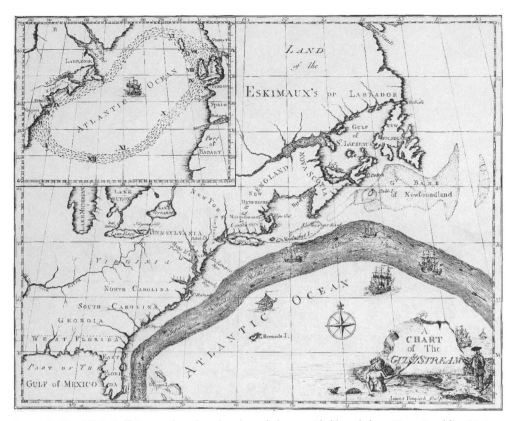

Benjamin Franklin's Gulf Stream chart, based on knowledge provided by whalers. (From Franklin, "A Letter from Dr. Benjamin Franklin to Mr. Alphonse le Roy, Member of Several Academies, at Paris. Containing Sundry Maritime Observations," *Transactions of the American Philosophical Society* 2 [1786]: 294–329; used with the kind permission of the American Philosophical Society.)

the Gulf Stream on their northern route to the United States. American whalers, however, easily recognized the Gulf Stream as a distinct body of water. They had observed that whales avoided certain areas of the North Atlantic and noticed that those waters had a different color and temperature. Merchant captains had no reason to notice these features, much less to record them systematically. Whalers kept detailed records of where and when they caught whales, to enable them to locate profitable fishing grounds in the future. This kind of working knowledge rarely came to the attention of the rest of the world, but in this case Franklin published a series of charts of the Gulf Stream derived from a sketch by Folger. A 1786 version of the chart served as a basis for subsequent charts of the Gulf Stream, and was not superseded until 1832.[5]

Besides providing information about winds and currents, whalers were

Whalers attacking a right whale, the prey preferred by the whale industry. When the numbers of these whales declined, the industry turned to sperm whaling, first in the Atlantic and soon after in the Pacific. Whalers formed the vanguard of European and American explorers in the Pacific, and also brought back the first observations about the ocean's depths. (Published by Currier & Ives, between 1856 and 1907; Library of Congress, Prints & Photographs Division, LC-USZ62–5097, Digital ID: cph 3b49659.)

the first population to experience the vast third dimension of the ocean as part of their routine work. Until the early nineteenth century, whaling was conducted from bases on shore. Whalers waited offshore on small vessels for "right" whales, so called for their attractive qualities of swimming near shore, yielding plenty of oil, and floating when dead. Whalers harpooned these whales, then towed the carcasses to shore to cut away and render the blubber. In the eighteenth century, as whale populations close to land declined, whalers boarded larger vessels for longer voyages. They still had to return to port for rendering until the introduction of tryworks, a furnace aboard ship in which a kettle could be heated to render blubber, severed the link to land. Prompted by declining numbers of other whales, whalers' shift to sperm whales was also spurred by the superior quality of the oil produced from their blubber and their production of a valuable waxy substance called spermaceti, used to make candles that burned with a brilliant, smokeless flame. Sperm whaling began in the Atlantic in the eighteenth century, with British whalers working near the Arctic and Americans in the southern Atlantic. The American Revolutionary War decimated the American whaling industry, and the hostilities of 1812 to 1814 depleted both fleets. Between those wars, American whaling recovered; then after 1815 Americans dominated sperm whaling, which entered a long period of expansion lasting until midcentury.[6]

As whalers became acquainted with their majestic and terrifying new prey, they learned about its habits. Sperm whales ranged farther from land, and in deeper water, than whales hunted previously. Stories circulated through the maritime world, each more fantastic than the last, about the strange behavior of these enormous creatures. One tale told of a whale killed in the Pacific in whose blubber was embedded a harpoon of a vessel then working in the Atlantic. Proponents of Arctic exploration and the search for the Northwest Passage used this story to argue for support of exploring expeditions.[7] Such stories commanded attention even when their veracity was suspect.

Expedition planners, including the organizers of the U.S. Exploring Expedition under Charles Wilkes, routinely consulted whalers. Over six months before the subsequent major American discovery voyage, the North Pacific Exploring Expedition, sailed, Lieutenant John Rodgers traveled to Boston to interview whaling captains about the nature of the seas he and the other members of the expedition planned to visit, because whalers were well acquainted with Pacific waters and islands. These captains suggested to Rodgers a study of the seabed to search for evidence of what whales ate. They believed that explorers could locate grounds where whales lived by analyzing bottom sediments.[8]

Not only did whalers provide navigational information but their methods of capture, using small boats and oars, put them in close proximity to their prey. Their accounts report on whale behavior, especially interactions between mothers and calves that whalers could exploit to capture both together.[9] Whalers also contributed some of the first observations about the great depths. Sperm whales could stay down for hours. When harpooned, they dove straight down, or sounded, taking an astonishing length of line with them. Whalemen surmised that whales could survive the pressure at great depths. Scientific experts, having concluded that life could not exist below about 300 fathoms, rejected the implications of such stories. But the tales persisted, and proliferated, as sperm whaling boomed. By midcentury they were hard to ignore.

Among scientifically inclined men who took whalers seriously was the famous but controversial American naval officer Matthew Fontaine Maury, whose career until the start of the Civil War was intertwined with the midcentury discovery and exploration of the sea. Like Franklin in the previous century, Maury bridged the gulf between the maritime world and the polite society of men of science by taking seriously the oceanic observations of whalers and other mariners. He and a few other scientifically inclined writers incorporated sea stories into their scientific works and narratives. Far from serving merely as diversion or entertainment, these stories provided compelling evidence to support scientific arguments.

Prevented by a leg injury from serving at sea, Maury reshaped his career. He set out to integrate science into naval practice and education by demonstrating its practical utility. In 1842 he was appointed director of the Navy's Depot of Charts and Instruments and soon thereafter he was made head of the newly formed Naval Observatory. At the depot he found a collection of logbooks that he used to compile information on winds and currents in the Atlantic for different months of the year. Mariners welcomed his wind and current charts because they provided information that shaved time off sea journeys.[10]

Extending this work, Maury began to collect logbooks and observations not only from naval officers and merchant shipping captains, but from whalers as well. By the early 1850s he had added to his repertoire a series of whale charts that he created by compiling observations of whales, including captures. He published these charts, along with information about the distribution and habits of economically valuable whale species, in his *Explanations and Sailing Directions to Accompany the Wind and Current Charts.*[11]

In this unlikely venue, Maury presented whalers' observations about marine life. He quoted Captain Francis Post, who described connecting three or four 225-fathom lines to prevent harpooned whales from escaping. Another captain wrote to Maury that he had seen whales sound deep enough to take 1,050 fathoms of line from the boats. These mariners ridiculed scientists for making absurd mistakes in describing whales "that are found in regions w[h]ere [the scientists] never venture themselves." They also dismissed scientists' musings about the effects of pressure on sea life in the face of the fact, which they personally witnessed at sea, that whales, even "small fry," could bear the pressure at more than 100 fathoms. Maury, whose preoccupation with the sea reflected a blend of commercial and intellectual interests, argued for the value of deep-sea soundings to promote the whaling industry. He hoped that deep-sea research would reveal how deep whales dove for food, whether they fed on the bottom, and whether they could live only in waters over a certain depth. These answers, he insisted, would ultimately increase "the wealth drawn out of the deep and conveyed . . . annually to the shores of America."[12]

A few British naturalists interested in the oceans also relied on the testimony of whalers, which they employed to build arguments about life at depths. Edward Forbes, the pioneer marine zoologist who inspired dozens of naturalists to dredge the British seas, praised the "early adventurers in the whale fisheries" for their careful observation of the characteristics and habits not only of their prey, but also of "the more striking among the minute organisms vivifying the polar waters." In his 1859 book, Forbes relied on Captain William Scoresby, a whaling captain and Arctic navigator whose friendship with Sir Joseph Banks encouraged him to study Arctic geography and natural history while working in the waters off Greenland. Forbes took seriously Scoresby's observation that sperm whales ate jellyfish which inhabited very deep water. Forbes also turned to the deep-sea fisheries, those pursuing sharks, sea perch, the ling, and the tusk, not only for information about these species, but also for "specimens worthy of national museums" such as the "curious creatures which cling to [the fishermen's] lines when they are engaged in fishing for the ling."[13]

George Wallich, the naturalist on HMS *Bulldog*, bolstered his argument that animals could live at great pressure with a story of Scoresby's about a whale dragging a boat down to such a depth that the wood from the boat sank like a stone not just when it was recovered, but for a year afterward. Wallich declared, "Here, then, it would appear that the whale must have descended as deep as the boat, or probably deeper, inasmuch as the latter was attached to it by the whale-line and harpoon." He continued, "Other instances have at various times been recorded of harpooned whales having dived down vertically to immense depths, and in some cases, having been killed by coming in forcible contact with the bottom!" Even as late as the *Challenger* expedition, naturalists such as John Murray sought out whalers for information about the areas where they worked or unfamiliar species they encountered.[14]

Whalers' "fish stories," then, joined explorers' narratives and hydrographers' reports to provide the land-bound scientific community with information about the deep sea. The first generation of scientific books about the ocean were indebted to the observations of people who worked at sea. In these volumes, whalers' stories and specimens were presented as reliable testimony. Whalers also contributed technology to early ocean investigation in the form of tapered lines designed to combat rope breakage when sperm whales swam away or dived deep after being harpooned.[15] By introducing scientists to the entirely unfamiliar world below the waves, these mariners opened the study of the depths.

Even when whalers' stories were not embraced as evidence, they served as incentives for scientific investigation of the depths. Whaling provided concrete commercial and political incentives to study the open ocean and the deep sea. Maury often invoked whaling as a significant economic rationale for studying the sea, equally important as investigating winds and currents for the benefit of shipping and passenger travel. Men of science such as Louis Agassiz likewise promoted ocean exploration, "not only in a scientific point of view but particularly and for the interest of our whale fisheries."[16] That exploration, when it happened, was spurred by economic and nationalistic motives and incorporated science as an integral activity.

THE NEW SCIENTIFIC style of exploration initiated by Captain James Cook profoundly shaped motives for plumbing the depths. This tradition emphasized systematic inventory of natural resources in addition to traditional geographic goals. Cook's enduring contribution was demonstrating that geographic discovery was most potent when followed up by detailed investigations to determine the botanical, mineral, and geographical assets of a place. This style of exploration emerged in response to pressing political and economic interests, including whaling, trading, colonial settlements, and new industrial concerns. Because colonial affairs often necessitated scientific advice, British exploring voyages began to carry a complement of civilian scientists. Although the inspiration for a particular excursion might derive primarily from one interested party or industry, planners ensured that expeditions served many simultaneous strategic, political, and economic interests. A ship sent out to locate a suitable port for whalers in the South Pacific would also carry a mineral surveyor to locate resources on the shores the expedition touched. Joseph Banks, who acted as the unofficial general director of British exploration during the last two decades of the eighteenth century, vigorously promoted Cook's style of exploration. New lands and seas were surveyed thoroughly, not only for utilitarian ends but also with an eye to increasing scientific knowledge.[17]

The concurrent emergence of national identities fueled this scientific style of exploration. National identities, which began to define European countries in the eighteenth century, were far from being natural, primary, or ahistorical. Territorial conquest was intimately linked with the formation of nations, while international conflict stimulated nationalism. Especially in peacetime, exploration transformed the globe into an arena for international competition over lands, resources, and markets.[18] Nationalism was a particularly strong force in Arctic and Antarctic exploration because of the sense that these were the last undiscovered places on earth. The same was true for the oceans.

England forged scientific-style exploration for imperial purposes, but the United States embraced this model because it suited the country's expansionist outlook. The period from the mid-eighteenth century through the nineteenth century in the United States was an age of discovery inspired by Cook's methods. The 1804–1806 Lewis and Clark expedition fit the model, since its goals were a blend of geographic discovery and the promotion of commerce and settlement. While discovery was ephemeral and accidental, exploration was deliberate and sustained. American exploration, especially after 1845, was characterized by a growing faith in science and a drive for precision manifested in precise mapping and charting. For overland exploration, soldier-engineers set out with the goal of producing maps and topographic surveys with an eye to settlement of the West. In this sort of exploration, government began to assume a more explicit and self-conscious role.[19]

U.S. ocean exploration was less constant than U.S. land exploration, but equally influenced by the model of scientific exploration. Between the years 1837 and 1860, the U.S. Navy sent out fifteen expeditions to all parts of the world. These expeditions fulfilled the international, seaward extension of American "manifest destiny" whereby Americans pursued commercial opportunities and sought to demonstrate the worth of their democratic ideals to the European world. Promoted mostly by amateur science enthusiasts, including naval officers, government officials, politicians, and private collectors, exploration offered proof of successful democratic science by virtue of the sheer quantity of its results. Voyages were spectacular, public events. Proponents of exploration wrested support from a government generally unwilling to spend federal money by appealing to the growing sense of frustration over America's "colonial," or dependent, status with respect to European science. Success in obtaining support for science in the 1840s and 1850s derived partly from a federal budget surplus, from which the army and navy especially benefited. The most famous of these naval expeditions, the U.S. Exploring Expedition of 1838–1842, commanded by Charles Wilkes, intended to make a mark for the United States vis-à-vis European exploration and science.[20]

In Britain, the legacy of Cook's voyages extended much farther than in the United States. The Admiralty maintained official policies that encouraged its surgeons and officers to devote their spare time to science. Scientific discoveries and publications were to be considered in decisions about promotion. Although in practice these policies were not always scrupulously adhered to, the British navy was friendlier than the American navy toward its scientifically inclined officers, and more likely to welcome nonmilitary scientists aboard naval vessels. Charles Wilkes fought hard to prevent civilian scientists from inclusion in his expedition party. But in Britain, especially during the many years Sir Francis Beaufort served as the Hydrographer of the Navy, the Admiralty earned a reputation for supporting science. Two widely acclaimed though atypical examples were Charles Darwin's voyage in the *Beagle* and Thomas Henry Huxley's in *Rattlesnake.* Darwin traveled by the personal invitation of Captain Fitzroy, while Huxley's marine research was not part of his official duties and was conducted with some difficulty. Indeed, the vehicles for scientific exploration were more often routine hydrographic survey cruises such as these rather than grand expeditions.[21]

During the first half of the nineteenth century, explorers were generally much less interested in the oceans themselves than in the seas and lands they visited. Scientifically inclined travelers such as Benjamin Franklin sometimes amused themselves during Atlantic crossings by capturing organisms to examine. Some expedition naturalists likewise relieved the tedium of long passages between landfalls by catching creatures in a surface net. While the *Beagle* sailed through the desolate region between Rio Plata and Patagonia, Charles Darwin "often towed astern a net made of bunting, and thus caught many curious animals." His *Voyage of the "Beagle"* contains few such references, however. The botanist Joseph Hooker, who sailed with Ross's 1839–1843 Antarctic expedition, examined surface fauna more systematically than previous explorers, and Huxley also devoted part of his *Rattlesnake* efforts to the study of marine invertebrates.[22]

Explorers made greater inroads into deep water hydrography than marine natural history. Until midcentury, though, few navigators, even hydrographers, cared how deep the water was as soon as their vessels were offshore. When making landfall on a strange coast, explorers sounded with a 200-fathom line, long enough to reach the continental shelf. Beyond that, depth was irrelevant, except in the vicinity of reported rocks or shoals whose position was not known precisely. Very rarely, explorers attempted to fathom the blue water. To illustrate how little was known about the ocean's depths before the nineteenth century, John Murray, the British naturalist who sailed with the *Challenger* expedition and became one of the founders of oceanography, repeated a story about Magellan, who, upon finding no bottom in the Pacific with a 200-fathom line, declared he had lo-

Deep-sea sounding from boats during Ross's Antarctic Expedition. (From James Clark Ross, *A Voyage of Discovery and Research in Southern and Antarctic Waters, 1839–1843*, vol. 2 [London, 1847], 355; courtesy of the National Oceanic and Atmospheric Administration, U.S. Department of Commerce.)

cated the deepest part of the ocean.[23] Whether true or not, the story emphasizes that mariners were more concerned, when sounding in deep water, with ruling out shallowness than with measuring exact depth.

The earliest systematic efforts to sound the deep ocean took place during early nineteenth-century polar explorations. Sir John Ross conducted soundings on his voyage to Baffin's Bay in 1817–1818, reporting a measurement of 1,000 fathoms. Two decades later, the U.S. Exploring Expedition under Wilkes tried sounding in deep Antarctic waters. About the same time, Ross's nephew, Sir James Clark Ross, departed on his Antarctic expedition determined to continue his uncle's work. Under Beaufort's direction, this Ross expedition fit well into the model of scientific exploration of the Arctic. In 1840, James Ross conducted deep-sea soundings in 2,425 and 2,677 fathoms. Twice, he got no bottom soundings with 4,000 fathoms of line out. These were the deepest soundings attempted to that date.[24]

Exploration became irresistible to scientists as a model for the organization and execution of ocean science. It placed them in the appealing role of intrepid explorers conquering nature by discovering its secrets. Knowledge about the oceans came to carry the same national importance as that gathered about foreign lands by earlier exploring expeditions. A well-known Swiss naturalist who immigrated to the United States, Louis Agassiz, drew this analogy explicitly, arguing that his investigations would "make us acquainted with the bottom of the ocean, as thoroughly as we are already with the surface of the earth." He concluded, "the results of this voyage will be as important for the increase of our knowledge of the characteristics of the sea, as the voyages of Captain Cook were, a century ago, for the improvement of navigation and geography."[25]

ALTHOUGH THE United States pushed westward in the middle decades of the nineteenth century, the nation also extended its seafaring frontier. The strong maritime focus of the early republic matured as American shipping swelled to compete with Britain's. By the late 1830s, the United States dominated passenger carriage and the transportation of high-value items, such as the mail, the news, specie, jewelry, and the latest fashions, between New York and Britain. Between that decade and 1860, the tonnage of American vessels increased by a factor of five. In support of this burgeoning industry, the U.S. Navy sponsored more than a dozen maritime exploring expeditions that promoted maritime science, technology, and commerce.[26]

One of these, the North Pacific Exploring Expedition, represented the first major effort by any nation to claim the ocean scientifically. The expedition sailed in 1853, during the decade in which the ocean's greatest depths shifted into focus as an object of scientific scrutiny. Until the expedition returned in 1856, its members pursued their mission to survey "for naval and commercial purposes such parts of the Behring Straits, the North Pacific Ocean, and the China Seas, as are frequented by American whaleships, and by trading vessels in their routes between the United States and China." Unlike many contemporary expeditions, its naturalists and scientifically minded officers did not concentrate exclusively on coastlines, islands, and ports. Instead, John Mercer Brooke, inventor of a detachable sounding device for recovering bottom samples, conducted the earliest confirmed deep-sea soundings in the Pacific, recovering the first bottom samples from that ocean's great depths. The expedition's naturalist, William Stimpson, a promising marine zoologist and probably the most experienced naturalist-dredger in the country, focused on organisms never previously studied by North Pacific explorers, marine invertebrates.[27]

The fact that the expedition's primary scientific focus was the study of the ocean was widely advertised. Stimpson explained to James D. Dana, "As the Expedition had for its object marine explorations and surveys only, no opportunity was afforded for inland research; so that fishes and invertebrates—the chief inhabitants of the sea—form the mass of the collections."[28] In part, the expedition's goals and accomplishments were guided by the surge of interest in the sea: naturalists' growing fascination with bizarre marine forms; the technological enthusiasm for plumbing ever-greater depths; and the background of commercial and political interest in the sea due to the tremendous profits reaped from shipping, whaling, sealing, and fishing. Japanese restrictions against foreigners coming ashore heightened official interest in organisms accessible from boats. The decision to concentrate the expedition's official scientific attention on marine invertebrates exemplifies the intersection of nationalism and science; the planners built into the expedition a guarantee that Americans would receive credit for discovering and naming hundreds of species.

The expedition marshaled virtually all practical and theoretical expertise available in the United States regarding deep-sea investigation, both biological and physical. Brooke and Stimpson were both selected for the voyage because of their unique expertise in deep-sea sounding and natural-history dredging, respectively. The expedition's chief officers were chosen for their exploring and hydrographic experience. Commander Cadwallader Ringgold had previously commanded one of the ships of the Wilkes expedition. Lieutenant John Rodgers, who was promoted to expedition commander after Ringgold left for health reasons, had worked for the U.S. Coast Survey from 1839 to 1842, and again in 1849, doing hydrographic surveying, including deep-sea sounding. Behind the scenes, Matthew Fontaine Maury wielded his considerable influence to promote the hydrographic goals of the expedition, including Brooke's assignment to the party, and to add a geographic focus on the Bering Strait area.[29]

To augment the experience of the expedition party, members sought additional advice from outside experts. Stimpson turned to Spencer Baird and Louis Agassiz, both of whom had acted as mentors and teachers to him. Agassiz, in fact, had recommended Stimpson to Navy Secretary John Pendleton Kennedy for the job. Rodgers also consulted Baird to discuss the best methods of conducting deep-sea soundings in order to get specimens of bottom sediment. The expedition's botanist, Charles Wright, asked Professor Jacob Whitman Bailey, the microscopist who examined ocean floor deposits collected by Maury and the U.S. Coast Survey, for advice on collecting and studying marine algae. Finally, in preparation for the expedition, Maury put Rodgers in touch with captains of whaling and merchant ships who worked in the ill-charted areas that the expedition would survey.[30]

The personnel assembled for the North Pacific Expedition represented the state of the art in knowledge of the ocean and its creatures. They were the vanguard of those Americans attempting to learn the secrets of the sea. Their accomplishments not only added new species and depth measurements that reflected well on their country, but also led to improved collecting techniques and practices.

At the time he sailed with the expedition, Stimpson ranked among the most skilled dredgers in the United States. He perfected his dredging technique under the tutelage of Louis Agassiz, who enjoyed generous patronage from Coast Survey director Alexander Dallas Bache, including the use of Coast Survey vessels. Stimpson used an array of methods traditionally employed by naturalists. While he was ashore, beachcombing at low tide was a favorite activity. Animals could also be found among floating sargassum weed and on or near the occasional floating casks and logs that the ship encountered. Once, flying fish landed on deck, only to be promptly preserved. Most frequently, Stimpson deployed tow nets, dip nets, large and small dredges, fishing lines, seines, and trawls. Early in the voyage, he made frequent use of tow nets and dip nets, focusing on pelagic fauna, including the phosphorescent animals that rose toward the surface at night. Later, he began to deploy small dredges to collect fauna from the bottom, mostly in harbors from the ship's boats, but occasionally from the flagship, the *Vincennes,* itself.[31]

Dredging from the flagship required a set of skills totally different from those needed to dredge from a small rowboat. Stimpson appears to have been among the first naturalists to apply dredging techniques on such a big vessel. On the *Vincennes,* Stimpson quickly learned that dredging was impossible if the ship was sailing fast and succeeded only when it moved quite slowly. At first, he dredged mostly when the ship was becalmed and barely drifting along. The first time he tried dredging with a large dredge from the ship, two hauls were successful. The third took so much sand that he recovered the dredge with difficulty, only to find the frame damaged. Later he thought of attempting to dredge when the *Vincennes* tacked, at which point the ship headed into and across the wind, thus slowing down dramatically from sailing speed before picking up speed again on the opposite tack.[32]

Stimpson began the North Pacific Exploring Expedition with some field experience and much advice. He left it with not only an awareness of the importance of understanding the distribution of marine life, but also a

A sketch by William Stimpson of a crab specimen collected during the North Pacific Exploring Expedition. (Smithsonian Institution Archives, RU 7253, North Pacific Exploring Expedition, Box 2, folder 5.)

vastly enhanced skill at manipulating dredges and other collecting equipment. His attention to native fishermen's instruments and methods enhanced his collecting ability. He improved his dredging method and learned to collect from deeper and deeper water simply by spending a lot time dredging, varying the depth, type of bottom, vessel, and speed. If he kept a record of his dredging, it has been lost, but his journal entries reveal that, while he worked in depths of 10 to 20 fathoms early in the voyage, toward the end he routinely reached 30 to 50 fathoms. Deeper attempts were risky. Once, Stimpson lost a large dredge in 50 fathoms despite using a six-inch (in diameter) hawser, a huge line, usually for docking or anchoring, because, "the frame catching on the bottom, the handles were torn out and the machine consequently lost."[33] When Stimpson returned to the United States, he shared what he had learned about deep-sea collecting on the expedition with American and British naturalists.

John Mercer Brooke also contributed to the techniques of deep-sea science. In 1853, Brooke received orders to report as the lieutenant for the tender *Fenimore Cooper,* which would accompany the expedition. Lieutenant Samuel P. Lee, under whom Brooke had served in the Coast Survey, recommended that Brooke be reassigned to the USS *Vincennes,* arguing that Brooke was too valuable an officer not to serve on the flagship. While working for the Survey, Brooke had learned how to obtain bottom sediment, one of its regular surveying practices since the 1840s. Later he was ordered to the U.S. Naval Observatory, where he compiled information, prepared charts, and received scientific training in hydrography, astronomy, and meteorology. Brooke was unhappy at the loss of command of the *Fenimore Cooper*, because he recognized that the navy rewarded only traditional duties, not scientific ones. Because of his scientific training at the Naval Observatory, though, Brooke was designated the official astronomer of the expedition and put in charge of the chronometers and other instruments. He was the only naval officer assigned purely scientific duties, although he still carried responsibility for standing watch and conducting hydrographic work.[34]

Having expressed to Maury "much interest in experiments for deep-sea sounding," Brooke carried out a series of sounding experiments during the course of the expedition. He determined that sounding from a boat yielded better results than sounding from the ship. In his first several attempts, he learned that the tallow for arming the deep-sea lead needed to be more viscous than that used when plumbing navigational depths. Finally, experiments with different kinds of line convinced him that heavy hemp rope was useless for deep soundings and lighter line was needed instead.[35]

Because the *Vincennes* was a survey ship, often working near land and in ill-charted areas, its crew sounded frequently. Their caution was well founded, since the ship had several close shaves with uncharted rocks. Many *Vincennes* soundings were in the same ranges that Stimpson dredged: up to about 50 fathoms, but more often under 30. Depths of 500 and 650 fathoms were not uncommon, though. Brooke tested his newly devised sounder when the *Vincennes* sailed over blue water. He discovered that deep soundings required considerable skill, which he developed only gradually. His early attempts included many failures and "no bottom" measurements, meaning that either the lead returned without any sediment to confirm it had struck bottom or the strain of hauling in broke the line so that the lead was lost. Brooke recorded depths between 2,000 and 3,000 fathoms, and once even 7,700 fathoms. Several months into the expedition, near the Solomon Islands, Brooke pronounced himself "exceedingly well pleased" with the successful retrieval of a bottom sample from 2,150 fathoms using his sounding device. Another achievement was a sounding the following year that recovered bottom from 2,700 fathoms. Brooke's deep-sea soundings were of zoological as well as hydrographic significance, because Stimpson microscopically examined sediment samples collected from the center of a quill Brooke attached to the sounding apparatus. In doing so, Stimpson became the first naturalist to study fresh, unpreserved bottom sediments at sea.[36]

Brooke's sounding experiments contributed to ongoing American efforts to solve the problems of sounding in the deep ocean. They also provided the only available information on the depths of the Pacific at a time when all other efforts focused on the Atlantic. Although Brooke's Pacific soundings represent some of the earliest accurate deep-sea soundings, they were not acknowledged as such by British observers. Charles Wyville Thomson assigned that honor to HMS *Porcupine*'s 2,435-fathom sounding, which he described as "probably the deepest sounding which had been taken up to that time which was perfectly reliable."[37] It is possible that Thomson never knew about Brooke's soundings or that he judged them inaccurate or discounted them for lack of published information.

Although important historically because it demonstrates early American commitment to studying the depths, the North Pacific Exploring Expedition went unheralded by midcentury ocean scientists. It has also been largely forgotten by historians, owing to the dearth of records and specimens documenting the expedition's achievements. No official narrative of the voyage was ever produced, a departure from the usual practice of government publication of such an official record. Nor was the expedition remembered for its scientific accomplishments because, although papers appeared in scientific journals describing some new species, no official volume of scientific results was published. After the expedition, Commander John Rodgers went on shore duty to oversee the preparation of reports and charts. With the outbreak of the Civil War, he was reassigned to the Mississippi River flotilla. After the war, nothing was published despite the fact that the narrative and scientific reports had been completed. In addition, most specimens in the principal collection, the marine invertebrates, were lost in the Great Chicago Fire of 1871.[38]

Although not well remembered by historians, the North Pacific Exploring Expedition represents an important benchmark for the history of ocean science. That the official scientific reports of the voyage were never published indicates, among other things, that the government was, perhaps, not entirely convinced that the collection of "creeping things" merited federal expenditure.[39] Failure to publish also reflected the fact that interest in the deep ocean did not yet form a coherent research agenda the way it would just a few years later, after submarine telegraph interests became more powerful. At this point, each group of investigators sought to learn about the sea for its own purposes: Naturalists extended their study of particular families or orders of marine forms, while hydrographers sought recognition for accomplishing deeper and deeper soundings. These impulses to reveal and understand the depths fueled curiosity about a place that was just beginning to exist in the midcentury imagination. Expeditions like the North Pacific added some knowledge of the mysterious domain of the deep sea and raised innumerable new questions.

Most important, this and other expeditions that featured dredging or deep-sea sounding produced the tangible benefit of techniques and technologies that allowed investigators to reach ever-greater depths. The individual collecting experiences of North Pacific expedition members were not entirely lost with the failure to publish expedition results. American hydrographers employed Brooke's sounder in the Atlantic from the time he invented it in 1854.[40] The British Hydrographic Office also adopted Brooke's sounder and, from the mid-1850s on, a cadre of officers particularly interested in deep-sea sounding adapted it and used it widely.

Stimpson devoted post-expedition years to getting his scientific results ready for publication. He continued his own dredging, even during the Civil War, once undertaking a cruise in spite of rumors that a privateer was operating in the area. After he got a job as director of the Chicago Academy of Sciences, he tried deep-water dredging in Lake Michigan, although with sparse results. Through his correspondence during the North Pacific expedition, as well as later conversations, letters, and dredging trips, Stimpson related his knowledge of how best to collect with dredges and trawls to interested naturalist friends. In 1860, for example, he took a dredging trip to South Carolina with Theodore Gill. Two years later, he spent several months dredging off the west coast of England with members of the British Association for the Advancement of Science (BAAS) dredging committees. The zoologist and Coast Survey officer William Healey Dall consulted Stimpson while planning fieldwork in Alaska, and the two agreed that Dall should visit Chicago on his way west. Stimpson explained, "I want to talk over the general subject of dredging with you . . . I can give you hints etc much better in conversation than in writing."[41]

Spencer Fullerton Baird, who as assistant secretary of the Smithsonian Institution was a "collector of collectors," became an important beneficiary of Stimpson's dredging experience. Through him, numerous naturalists who visited, worked, or studied at the Smithsonian also benefited secondhand from Stimpson's experience. As he had for Stimpson and Rodgers, Baird continued to serve as a consultant for those interested in deep-water collecting, acting, as he often did, as a general clearinghouse for information about natural-history practices within a close-knit group of young naturalists he fostered and mentored.[42]

The North Pacific Exploring Expedition assembled, and expanded, available expertise in measuring and collecting from the deep sea. Its members made significant contributions to the practice of nineteenth-century American ocean science as well as to the emerging scientific investigation of the depths. The skill Stimpson acquired—of dredging directly from large ships rather than small boats—would not be employed again until naturalists began to find places for themselves on naval and hydrographic ships in the late 1860s. But the handful of accurate deep-sea soundings made during the expedition presaged the flurry of hydrographic activity in the late 1850s devoted to transforming these infrequent and uncertain experiments into routine and reliable measurements. In addition, the sheer number of marine species from zoologically unknown Pacific waters that Stimpson collected, described, and named represented an important American achievement in marine zoology, one that did not go unnoticed across the Atlantic. Most of all, the expedition exemplified the cultural association developing between the ocean's depths and the tradition of scientific, geographic exploration previously focused exclusively on land.

To the young William Stimpson, the North Pacific Exploring Expedition promised much more than the opportunity to collect some specimens. The emerging cultural idea of exploration as a means to study the sea manifested itself on a personal level in the career ambitions of young scientists, for whom voyages of exploration became an acknowledged route for seeking their intellectual fortunes. Increasingly, returning with important collections and new ideas not only opened doors within the scientific community, but often produced a job that allowed this first generation of professional zoologists to pursue their own research while earning a living. Although the reality did not always match the expectations that naturalists brought with them to sea, the expectations themselves were commonplace and strong. Stimpson aimed to follow in Edward Forbes's path. Before Forbes accompanied the 1841 cruise of HMS *Beacon* to survey the coastal waters of Greece and Turkey, he occasionally gave lectures or taught, but relied on financial support from his father to pursue science. During the cruise, on which he was the official naturalist, he dredged extensively and studied marine animals. Afterward he was able to secure employment in science, at first cobbling together a teaching position in botany with the curatorship of the Geological Society of London, later becoming a paleontologist for the Geological Survey under Henry de la Beche, and finally occupying the prestigious Regius Chair of Natural History at the Universtiy of Edinburgh.[43]

Because the ocean represented a completely unexplored environment, naturalists intending to study marine fauna entertained particularly strong preconceptions about the likely career outcome of their voyages. The stories of how men such as Joseph Hooker, Thomas Henry Huxley, and Charles Darwin got their start aboard British vessels sent to explore and survey distant lands and waters seemed to testify to the efficacy of ocean voyages as an entry into the scientific community.

In Britain, survey and exploration missions were placed under the auspices of the Hydrographic Office after the Napoleonic War. Under Hydrographer Francis Beaufort, the Hydrographic Office worked with the Royal Society and Royal Geographical Society to create a place for natural history in naval survey ships by assigning collecting duties to medical officers. Indeed, the naval medical office encouraged collecting. Having the qualifications of a surgeon made a naval commission, even primarily as a naturalist, easier to acquire. But in spite of the institutional tolerance, even encouragement, of natural history, the promise of promotion based on scientific work was not often realized.[44] In fact, the naval naturalists Joseph Hooker and Thomas Huxley concluded sadly that scientists who wished to do serious scholarly work would have to leave the service. Such insight came only with experience, however. To aspiring naturalists, these models of success beckoned, and many sought berths on naval vessels.

In the United States, the appeal of exploration voyages as vehicles for scientific advancement was also strong. The tradition of including scientists on exploring missions dated from the Lewis and Clark expedition and extended through the naval exploring expeditions of the first half of the century.[45] Preoccupied with measuring up to European scientific standards, American men of science adopted the naturalist-as-explorer model. Because much American natural history concentrated on native species, scientist-explorers traveled across land with the Geological Survey more often than they boarded naval ships to collect in foreign ports. Yet scientists who focused on marine fauna, including William Stimpson, Louis and Alexander Agassiz, and William Healey Dall, embarked with the same expectations as their British counterparts.

The British microscopist George Wallich provides a clear example of an ocean scientist who entertained firm and detailed expectations that a voyage of exploration would open a career in science to him. He went to sea, he explained, "for the acquirement of the reputation I covet." Wallich sailed with HMS *Bulldog* as a microscopist on the 1860 expedition to explore a more northern route for the Atlantic submarine telegraph cable than the one used in the 1857 and 1858 cable-laying attempts. Wallich's father, Nathaniel, was superintendent of the Calcutta Botanical Gardens and a good friend of the famous botanist and director of the Royal Botanic Gardens at Kew, Sir William Hooker. George was friendly with Hooker's son Joseph and knew about young Hooker's travels to the Antarctic with the Ross expedition and his subsequent scientific career. Hooker spent a few years following his return living with his parents and working up his results. He also gave lectures at Edinburgh as part of a failed attempt to get the natural-history chair there. Three years later, he secured a position as paleobiologist with the Geological Survey, which tided him over until his appointment as assistant director at Kew.[46]

For Hooker, an ocean voyage had always been part of his plan to establish himself as a scientist. The year before his appointment at Kew, he confided to Charles Darwin, "From my earliest childhood, I nourished and cherished the desire to make a credible journey in a new country, and write such a respectable account of its natural features as should give me a niche amongst the scientific explorers of the globe I inhabit, and hand my name down as a useful contribution of original matter." When Wallich's opportunity to participate in such an expedition came, he remembered his friend's experience. He wrote to Hooker seeking advice about his upcoming adventure and asked him to study the Greenland plants he planned to collect on the trip. Wallich admitted to him that "the opportunity is one I have all along had in contemplation and in order to take advantage of it . . . I worked at these microscopic affairs of mine."[47]

Wallich actively pursued a berth on the *Bulldog* through repeated visits to Professor David T. Ansted, Huxley, and Sir Roderick Murchison, seeking their support. He persistently lobbied Hydrographer Captain John Washington, arguing for the need to send a naturalist in addition to the regular medical officers of the *Bulldog*. Finally, he presented his case to the "leading men in the Telegraph Enterprise." Wallich secured impeccable recommendations for the position of naval naturalist on the *Bulldog*, all from men who thought of expeditions as unparalleled opportunities for promising scientists. Huxley had himself parlayed his years as assistant surgeon on HMS *Rattlesnake* into an established position in the Government School of Mines, having obtained the lectureship vacated by Edward Forbes, who left to take the Edinburgh natural-history chair. Murchison made ample use of the natural resources of colonies and scientifically unexplored places to establish his career. Wallich embarked with the assumption that this voyage would do for him what Hooker's, Huxley's, and, most famously, Darwin's had done for them: establish their reputations firmly (and positively) in the British scientific community.[48]

Hooker's and Huxley's experiences helped them secure employment that allowed them to combine scientific research with a good enough living to support a family. This last goal was especially important for Huxley, who disembarked impatient to marry his Australian fiancée of four years. Wallich, too, needed to earn a living. He started his career in HM Indian Army at the age of twenty-three. Nineteen years later, having acquired a wife and two children (the first of eight), he was invalided out and returned to England to tackle the problem of finding employment suitable for his station and his growing family.[49]

Wallich's background predisposed him to try to make a career in science. As early as 1849, he began collecting and studying marine natural history, corresponding with a Professor Bell for advice and help with species identification. Wallich's interest in marine fauna dates at least from 1850 and 1851, when he made surface collections and sketched marine organisms to pass the time during voyages to and from England. During his final return voyage from India in 1857, he made microscope slides of microorganisms found in the guts of salpae, free-swimming oceanic tunicates. In his *Bulldog* journal, he reflected on the reason he chose marine zoology as his scientific focus, "The other branches of Natural History seemed to me to be more or less fully occupied by men fully competent to deal with them, whilst in the one selected by me the field might almost be said to be untrodden." Wallich was so confident that his experience on the *Bulldog* offered an avenue to employment that he volunteered to sail without pay. The Admiralty did award him a naturalist's pay of £300 per year, but he had to dig into his own pocket for collecting gear as well as for occasional hiring of local boats and fishermen for inshore dredging. In the midst of his work at sea, while microscopically examining bottom specimens, Wallich worried about the ramifications of these collections for his scientific career: "The soundings of course afforded me some interesting matter, but not of the varied character I coveted."[50]

In spite of fears and doubts that plagued him throughout the voyage, his immediate post-expedition situation seemed promising. About a month after he returned, he lectured at the Royal Institution to an appreciative audience of Murchison, Michael Faraday, and "the whole of the London celebrities of the scientific world." Wallich proudly chronicled the moment Faraday fetched him a cup of tea: "Olympian Jove bending down to Pan." As Huxley had, he received a publication grant from the Admiralty, which he used to publish his book, *The North Atlantic Sea-Bed,* in 1862. He also delivered a series of lectures in Hull. But when these ended four months after the expedition returned, he expressed frustration at receiving only thanks for all his hard work, especially in light of his pressing need to secure employment.[51]

Ultimately, Wallich failed to make a place for himself in the London scientific world. Although nominated for Fellowship in the Royal Society in 1864 by more than half a dozen respected men of science, including Huxley, Murchison, and Sir Richard Owen, he was not elected that year, nor did Owen remember to renominate him the next year. In 1867, Wallich approached Hooker for a renomination, but was rebuffed. By 1866, Wallich had given up and set up shop as a photographer, explaining to Hooker that, "I have after long & mature consideration engaged in my new occupation in the hope that I may thereby be enabled to benefit my wife and children . . . I have a good chance of success. At all events more than I ever could venture to hope for in that muddy sea of science."[52]

There are several possible reasons for Wallich's failure to transform his *Bulldog* experience and collections into a career. Perhaps voyages of exploration no longer opened doors so easily in a community increasingly composed of professional scientists. Or maybe the emerging community of ocean scientists had settled upon other requirements for admission. On a personal level, Wallich alienated his early supporters. He was quick to anger and to accuse others of plagiarizing his work. In 1877, he attempted to force the Royal Society to sanction William Benjamin Carpenter and Charles Wyville Thomson on grounds of plagiarism. Never during his life did he feel as though he received the attention he deserved for his scientific discoveries and his inventions. Immediately after the *Bulldog* expedition, he fought bitterly with an impassive Admiralty over a sounding machine he helped to develop during the voyage. His prickly, oversensitive nature, as well as his verbose and cumbersome prose style, greatly exacerbated his disagreements as well as his feelings that he was not adequately recognized.[53] Ultimately, Wallich failed to reap what he believed to be the deserved outcome of a sea voyage: recognition by the scientific community, cemented by a secure job.

The explorer-naturalist model was attractive not just to new naturalists, such as Wallich, seeking recognition for their work, but also to established men of science. William Carpenter, who worked with Charles Wyville Thomson in the late 1860s to arrange for dredging expeditions on HMS *Lightning* and *Porcupine*, lobbied strenuously for the position of scientific director of *Challenger*, which ultimately went to the younger Thomson. He wrote to Sir Edward Sabine, then president of the Royal Society, reviewing his career and emphasizing how hard he had to work at lecturing and popular writing in order to pursue original research. He declared that devoting the rest of his life to the scientific work he had begun on *Lightning* and *Porcupine* would be "the highest objective of my ambition. . . . It has been 'the dream of my life' to engage in such researches in Tropical

Seas." "When Huxley came home from his 'Rattlesnake' Voyage," Carpenter wrote, "I told him (as he remembers well) that I would have given almost anything for such an opportunity."[54]

The image of the intrepid, voyage-making naturalist also became familiar to a wider population. The nineteenth-century reading public met heroic naturalist-explorers through the popular genre of the travel narratives that proliferated in that century. In *Two Years before the Mast,* Richard Henry Dana wrote of meeting his former college professor who sailed as a passenger in order to collect specimens in California. The sailors "puzzled to know what to make of him" and his odd habits, especially his decision to "leave a Christian country . . . to pick up shells and stones."[55] Although leaders of the great exploring expeditions were revered as heroes, only a few of the men who sailed as surgeon-naturalists or naturalist-explorers garnered much personal fame. The efforts of most of them bolstered the political and cultural aims of their expeditions much more than their own personal scientific success. Nevertheless, from the perspective of the shore, ocean voyages seemed to promise an avenue for entering the scientific world.

OBSERVING, SAMPLING, mapping, naming, and charting provided nineteenth-century explorers and scientists with tangible ways to demonstrate both personal and national claims to the atmosphere, mountains, and polar regions, as well as the sea. Between 1840 and 1880, the ocean ceased being a wasteland and highway and was transformed into a destination, a frontier, an uncivilized place ripe for conquest and exploitation. Before its depths were explored, geophysical scientists ordered, graphed, and thereby controlled the sea's surface through studies of the earth's magnetism and tides.[56] The ocean's depths, discovered first by whalers in pursuit of commercially valuable sperm whales, became a proving ground for explorers such as Brooke who conducted singular, spectacular sounding attempts in previously unfathomable depths.

The act of inscribing a numerical depth on a chart offered a kind of personal glory and national prestige similar to that derived from discovering a new island, inlet, or anchorage, or finding and identifying a new species. Scientific or technological feats demonstrated cultural superiority as effectively as discovering and claiming geographically important or economically valuable places on earth. When the ocean was envisioned as a new frontier, routine hydrographic cruises, especially deep-sea sounding cruises, were heralded as national achievements.

A cartoon by Edward Forbes, (From Forbes, *The Natural History of the European Seas* [London: John van Voorst, 1859], viii.)

This definition of the ocean as an arena for national accomplishment and imperialist influence carried over into natural history as well. Edward Forbes's description of the sea bottom reveals this new attitude: "When a whole dredgeful of living creatures from the unexplored depths appeared, . . . it was as if we had alighted upon a city of unknown people. . . . And when, at close of day, our active labours over, we counted the bodies of the slain, or curiously watched the proceedings of those whom we had selected as prisoners and confined in crystal vases . . ." His language betrayed his starkly imperialist perspective as he concluded, "our feelings of exultation were as vivid, and surely as pardonable, as the triumphant satisfaction of some old Spanish 'Conquistador.'"[57]

The North Pacific Exploring Expedition prompted a measure of respect for American science from the European scientific community, which noticed American efforts to stake hydrographical and zoological claims on the ocean's depths. Professor Christian Gottfried Ehrenberg, the Berlin microscopist who examined Atlantic bottom samples for Maury, acknowledged American precedence in deep-sea investigation: "It is only the scientific efforts of the Americans, as yet, which furnish these [specimens] and it is especially Mr. Maury who animates Americans and other dwellers upon the earth to obtain knowledge of the sea." In the case of the North Pacific expedition, such observations accurately reflected American intention. Decisions to explore the ocean, to chart little-known islands and mainlands, and to pursue marine invertebrates shaped an undertaking that opened the new frontier of the ocean's depths to Americans. Even before the expedition departed, William Stimpson wrote to Commander Ringgold suggesting how publication of species lists and identifications should be managed in order to maximize credit to the expedition and, therefore, to the American Republic, "not only in the eyes of its inhabitants, but also in those European nations, by whom zoological research has been heretofore regarded with much greater respect than generally here."[58]

Indeed, exploration of the sea became a source of national pride for Americans. To Maury, sea floor examination was an appropriate exercise of naval resources and he was proud that, "the subject of deep sea soundings did not escape the attention of an enlightened government wisely administered." In 1853, he celebrated American naval leadership in improving deep-sea charting, observing that "no maritime nation has heretofore undertaken a systematic search for . . . dangers which render navigation uneasy, if not unsafe." The American navy not only offered immediately tangible benefits to commercial shipping, but also provided its officers with the means of "increasing the stock of human knowledge." Citing the government's "liberality and enlightenment beyond all praise," Maury pointed out that the implements, and also the responsibility, for conducting deep-sea soundings, "had been placed in the hands of American navy officers and in their hands alone." In 1856, the Secretary of the Navy wrote in the annual report, "I confess I felt some pride in having the science and naval genius of our own country to continue foremost in these great ocean surveys." The American government provided naval officers with the rare opportunity to contribute substantially to a new and burgeoning department of physical geography, that of the sea. So novel was the field, and so quintessentially American were the means to study the ocean, at least initially, that "the eyes of the scientific men of the world" fixed on the American officers engaged in deep-sea work, awaiting their results with great anticipation.[59]

Charting of coastal waters became an issue of national pride because of its tight links to commerce and security. The founding of the British Hydrographic Office in 1795 had been inspired in part by the desirability of England's gaining "her rightful place in chart-making among the maritime nations of the world." Before that time, French and Dutch charts were often better than British ones, despite the many British surveys of distant waters. The first Hydrographer of the Navy, Alexander Dalrymple, concentrated on compiling numerous existing charts. The Hydrographic Office began active surveying under his successor, Captain Thomas Hurd, at which time the office also made its charts available to the mercantile marine and the public. During the twenty-six years that Sir Francis Beaufort, Hurd's successor, served as Hydrographer, the office became a separate Admiralty department engaged in surveying activities all over the globe. By midcentury, the Hydrographic Office produced most of the charts used worldwide. Maury complained to the Secretary of the Navy that American ships had to rely on charts from other nations not just in distant seas, but also in American waters. He explained, using the capital city for emphasis, "The charts used by an American man of war when she enters the Chesapeake Bay, on her way to this city [Washington] are English and we are dependent upon the British Admiralty for them."[60]

The prospect of submarine telegraphy appeared to confirm this impulse to define the ocean and its depths in nationalistic terms. At the 1865 meeting of the British Association for the Advancement of Science, participants were told, "Such an enterprise as this [the Atlantic cable] reflects credit, not only upon the individual promoters of the undertaking, but also upon the nation itself."[61] Nationalism, fanned by the strong tradition of geographic exploration, strongly reinforced other motives that turned the attention of hydrographers and scientists seaward at midcentury.

chapter three

Soundings

Of what use is a bottom if it is out of sight, if it is two or three miles from the surface, and you are to be drowned so long before you get to it, though it were made of the same stuff with your native soil?

—Henry David Thoreau, *Cape Cod,* 1864

It was a sealed volume, abounding in knowledge and instruction that might be both useful and profitable to man. The seal which covered it was of rolling waves many thousand feet in thickness. Could it not be broken?

—Matthew F. Maury, *The Physical Geography of the Sea,* 1855

Efforts to reach beneath the "sealed volume" of the ocean's waters began at midcentury. Since the sea was unknowable except through the mediation of sounding devices, understanding of the sea floor depended heavily on instruments, skill, and assumptions about the action of sounding devices beneath the waves. By using the analogy of dry land geology, George Wallich pointed out the inadequacy of the methods available to midcentury hydrographers: "Suppose a person were to walk across the Welsh Mountains from one end to the other in a straight line or nearly so, and were blindfolded, to dig into the ground beneath his feet at certain intervals a cylinder like that of Brooke's machine. Could he by any possibility in this manner obtain an average sample of the soils or surface structure he would pass across? Most certainly not." Wallich's point appeared as ludicrous for geology in the midnineteenth century as it does today. Yet ocean scientists had no alternative to what Wallich characterized as "steeple chase fashion," or wildly irregular, sampling.[1]

Commercial opportunities attracted investigators to the ocean's third dimension. The first regular program of deep-sea sounding emerged from practical studies of winds and currents motivated by the growth of shipping and whaling, rather than from grand exploring expeditions. During the 1850s, American hydrographers working under the direction of Matthew Fontaine Maury undertook research cruises that had as their goal study of the ocean's greatest depths. Each individual sounding attempt during those early years was experimental, and the image of the shape of the sea floor that began to take shape changed dramatically several times within just the first few years of deep sounding.

Changing images of the sea floor reflected refined sounding methods and improved technology, to be sure, but they also measured shifting motives for investigating the sea. Before the commercial and strategic potential of submarine telegraphy drove investigations, the ocean floor appeared to hydrographers as a violent, rugged place. Later, with the commercial and scientific uses that midcentury engineers and scientists had in mind for the deep sea, its floor metamorphosed into a flat, quiescent environment safe for submarine telegraph cables.

NAVIGATION ESTABLISHED a commercial basis for early deep-sea investigation. Sounding had always provided important information to mariners, but increased chart-making activity in the nineteenth century reinforced navigators' reliance on knowledge about the depth of water and the composition of bottom sediments. Supported by a burgeoning maritime industry, American hydrographers pioneered deep-sea sounding while searching for aids and dangers to navigation.

As long as sailors have ventured to unknown seas, they have deployed lead lines to learn the contours of inshore waters. Measuring depth for navigational purposes was a relatively straightforward operation. A sailor, standing in the bow or working from a small boat, cast a lead attached to a line over the side and let it run through his hands. He counted the markings on the line that indicated how many fathoms had run out until he felt it hit bottom. Leads could be "armed" with grease or tallow so that they picked up a few grains of bottom sediment. As important navigational tools, sounding lines and leads were carried on all types of vessels. Explorers, or mariners sailing in unfamiliar or uncharted areas, sounded to prevent running aground or hitting rocks, to discover or rule out shallowness, not necessarily to measure precise depth.

Navigators deployed sounding gear either to locate themselves on existing charts or literally to feel their way in as they approached an unfamiliar coast. All sailors used soundings to locate themselves on their mental chart of local waters. A well-known folk tale about a Chesapeake Bay oyster fisherman explains how sounding assisted navigation. On a foggy day, the fisherman's crew took some chicken manure out to sea with them as a joke, and pressed it on the lead, which had been greased. When the captain inspected the lead, he smelled and tasted the "bottom," then leapt up, agitated, and cried out, "Luff up, boys! Something's wrong! We're in Mrs. Murphy's hen yard on Smith Island."[2] As navigators neared shore at night or in foggy conditions, they consulted charts marked with the nature of the bottom sediment—sand, rock, gravel, mud. Knowing the numerical depth and type of bottom often helped mariners ascertain their position at sea, and precise charting was crucial for safe maritime commercial activities.

To construct a hydrographic chart, surveyors not only carried out lines of sounding within a harbor or port, but also ran longer lines out to sea to record the approach to shore. Increasingly higher standards of surveying prompted hydrographers to carry lines of sounding farther and farther out to sea. By midcentury, instructions insisted that offshore surveying parties "should invariably keep the deep-sea lead going at regular intervals" of at least ten, twenty, or thirty miles. These orders stipulated that soundings be carried out to the edge of the continental shelf, whenever possible to the 100-fathom mark. In fact, "no opportunity should be missed" to sound in deep water, regardless of the prospects for reaching bottom.[3] For regular hydrographic purposes, though, the negative knowledge conveyed by "no bottom" soundings with a 1,000-fathom line was as valuable as actual depth measurements.

The sheer amount of chart-making activity increased dramatically in the nineteenth century in response to the growth of ocean shipping and passenger travel. The association between American maritime industry and scientific or technological achievements was older than the country in many respects. Early American science was largely based on the young nation's ties to the sea. In particular, Nathaniel Bowditch's mathematical accomplishments related closely to the practical needs of navigation, which he promoted vigorously.[4] During the nineteenth century, this link grew even stronger.

The close relationship between seafaring and the growth of American science emphasized the extent to which the country's maritime commercial needs drove the development of scientific institutions. Maritime industry not only helped finance westward expansion but also promoted the establishment of the Coast Survey, the Depot of Charts and Instruments, and the Nautical Almanac Office. These institutions supported a wide range of scientific research, encompassing meteorology, terrestrial magnetism, physics, astronomy, and natural history. From 1815 to 1860, the American merchant marine increased fourfold. As a result, the commercial interests of ocean shipping became a powerful ally to these institutions. In the 1840s and 1850s, when the army and navy became beneficiaries of unexpected surplus government funds, science related to seafaring activities received a significant share. Thus the alliance between promoters of American science and those with political and financial power in the great port cities helped create government institutions that supported scientific research.[5]

Two American institutions that undertook early scientific study of the ocean's depths were the U.S. Coast Survey and the Depot of Charts and Instruments. Congress authorized the Coast Survey in 1807, although it did not begin continuous operations until 1818. Shifts between civilian and naval leadership disrupted progress until 1843, when Alexander Dallas Bache assumed its directorship. Bache succeeded in building the organization into one of the premier surveying institutions in the world. The Survey's mandate included mapping the shoreline, islands, anchorages, and waters within twenty leagues of U.S. territory. Its responsibilities extended to examining banks or shoals such as Georges Bank, an important fishing ground, and to conducting hydrographic studies of coastal waters, including the Gulf Stream.[6]

The Depot of Charts and Instruments opened in 1830, but its first three directors concentrated on rating chronometers, maintaining navigational instruments, and creating an astronomical observatory. Not until Matthew Fontaine Maury took over as director in 1842 did the Depot devote its institutional resources to studying the sea. With the Survey's jurisdiction essentially limited to the coasts, Maury asserted naval right to investigate and chart offshore waters. Although both mariners and scientists recognized Coast Survey charts as of the highest quality, many observers, including Maury, felt that the Survey's exacting standards unnecessarily slowed production of useful charts. Maury therefore argued for the propriety of Depot involvement in publishing oceanic charts. He conducted original wind and current investigations to locate faster and safer sea routes. His practical orientation meant that the scientific and hydrographic research he conducted remained primarily maritime in its orientation, whereas Bache supported the broader work of professional scientists whose interests encompassed other fields in addition to ocean science.[7]

Maury's interest in studying winds, currents, and distribution of whales was rooted in the practical benefits he believed would accrue from such investigations. In spite of his attention to these surface phenomena, Maury was in the 1840s already developing an interest in the depths. In 1841, he urged the new National Institution for the Promotion of Science to encourage scientific exploration of the sea bottom.[8] By late in that decade he was touting the benefit of deep-sea soundings for safe navigation.

Although navigators generally felt safe traveling across deep water, they occasionally saw, or thought they saw, exposed rocks or shoal areas. Such deep-water hazards, or vigias, seemed especially dangerous because of their unexpected appearance in otherwise open seas. Vigias were frequently reported, although seldom found when searched for. After a fisherman or other mariner notified the authorities, hydrographers sent to investigate vigias attempted either to locate them precisely on the chart or to disprove their existence by confirmed soundings at and around the area in question, "for if they exist," as Maury explained, "you will probably find shoaler water in their vicinity." Clearing reported shoals and rocks off charts boosted commerce by opening new routes that navigators had previously dismissed as too dangerous. The political importance of investigating vigias was evidenced by the reports of the cruise of the U.S. surveying vessel *Dolphin* in 1850–51. Rather than emphasizing the expedition's deepest soundings, which represented its most impressive scientific and technical feats, Maury instead pointed to the long list of vigias whose existence the cruise disproved, noting the benefit for shipping of this accomplishment.[9]

IT IS DIFFICULT to express adequately human ignorance of the ocean's depths up to the midnineteenth century. In 1858, *Harper's Weekly* boasted that the North Atlantic seabed could be mapped with more accuracy than the interior of Africa or Australia.[10] This standard was hardly high. Before 1849, only a handful of efforts to fathom the depths had ever been made. Most of these were executed by explorers whose primary goal was geographic exploration of places more commercially or culturally interesting than the depths. Midcentury deep-sea sounding represented an entirely novel undertaking.

Initially, most knowledge about the depths was folkloric. Indeed, some people even in midcentury doubted that direct knowledge of the deep sea was possible. Many sailors believed that wrecked ships, bodies, or other objects thrown overboard did not fall to the bottom, but stopped and floated when they "found their depth." If this was true, objects would fall through the water only until the point at which the water's specific density equaled their own, at which depth they would float forever. Such ideas persisted far into the nineteenth century, embraced by some educated mariners and landsmen even after water was proved relatively incompressible.[11]

In 1857 a British navy lieutenant named Francis Higginson tried to warn naval officials and submarine telegraph promoters that sounding great depths was impossible, because the weight would stop falling long before it reached the bottom of the sea. His claim to expertise was based on forty years at sea as well as some experiments lowering cannons until they were displaced by denser water. Fortunately, he argued, this fact did not threaten the practicability of the Atlantic cable because it, too, would float deep below the surface. Higginson warned, however, that the cable company should rely on him for calculations and cable-laying advice because its current advisors labored under misapprehensions about the physical properties of seawater at great depths. Most midcentury hydrographers—and cable companies—rejected Higginson's ideas, but the nagging doubt that the sea could be conquered persisted, and resurfaced, with every major setback to the Atlantic submarine telegraph project.[12]

In this atmosphere of widespread misgivings, early attempts to study the depths were modest ones incorporated into an established research program. By the late 1840s, Maury had become well known and was respected by the naval and maritime communities in the United States and abroad for his textbook on navigation as well as his investigations of winds and currents. Maury's renown, as well as the economic and strategic value of his findings up to that time, convinced Congress in 1849 to authorize vessels to assist his investigations of the ocean and to test new sea routes.[13] The following year, Maury sent the U.S. schooner *Taney*, with Lieutenant. Joseph C. Walsh commanding, to the Atlantic with instructions to take hourly surface temperatures, temperatures at various depths, deep-sea soundings, and measurements of transparency, salinity, and specific gravity. Maury also instructed Walsh to search for vigias whenever he sailed in the vicinity of dangers marked on the charts he carried.

Most of all, Maury emphasized the uniqueness of the *Taney*'s mission: the fact that the ocean itself was now the object of investigation. For the first time ever, a vessel set sail neither to travel from shore to shore as fast as possible nor to survey a harbor or approach to land. Maury informed Walsh, "As the service on which you are engaged has for its object the making of observations and the collection of facts at sea, you will keep the *sea* during your absence as long as practicable." In addition, the *Taney*'s orders carefully stipulated that even deep soundings were to be made with "a view to reach bottom," a brand new priority to hydrographers previously concerned only with ruling out shallowness.[14] Before the *Taney*'s voyage, dallying at sea meant courting danger, boredom, or, for the merchant marine, loss of money. Even explorers searching for new sea routes sought the most direct path to land. With this modest beginning, the high seas themselves became the subject of investigation.

Unfortunately, the *Taney* proved dangerously unseaworthy, a poor choice for a ship ordered to remain at sea. Under Captain Walsh, the *Taney*'s crew did, however, sound "the greatest depth ever attained," 5,700 fathoms. This result startled hydrographers and geographers, who did not expect such vast depths. Most had assumed that the ocean's depth would approximately equal the height of land. Noting that the *Taney*'s work indicated that the Atlantic measured more than six and a half miles in the deepest parts, Maury marveled at the implication: "that the greatest depths of the ocean exceed the greatest elevations of the land by more than one mile at the least."[15]

Walsh experienced great difficulty in sounding. He and Maury had decided to use steel wire for these deep sounding experiments, because hemp lines would have to be incredibly thick to resist breaking. Indeed, as they made this choice, a British hydrographer, Captain Edward Barnett of HMS *Thunder* also experimented with iron wire for deep-sea sounding. But wire proved too breakable, as Walsh learned the hard way. His sounding wire broke at almost every sounding below 2,000 fathoms, including the 5,700-fathom measurement. Because that had been the *Taney's* first sounding attempt, Walsh lost a great quantity of wire almost immediately.[16]

Despite his difficulties, Walsh asserted complete confidence in the spectacular measurement of 5,700 fathoms. He recorded that the sounding line was perpendicular to the ocean's surface—or *"dead up and down"*—and stayed under the ship throughout, two important criteria for traditional soundings. No check slowed the line's descent, which was steady and uniform. These "carefully marked and measured" observations "prove that it could not have touched bottom before the break." Maury concurred with Walsh's assessment and agreed that Walsh's success demonstrated the feasibility of deep sounding. Nevertheless, he also noted, "It is very desirable to have this sounding verified."[17]

During the decade after the *Taney's* voyage, Maury promoted deep-sea sounding on special missions as well as routine naval cruises. For early deep-sea hydrographers, each measurement constituted an experiment in which every facet of sounding, from the type of line used to management of the crew's labor, had yet to be established. As Maury compiled depth measurements, he found vast discrepancies between sounding results from the same areas. These irregularities made hydrographers realize that they sometimes misinterpreted the moment that the sinker touched bottom. Although they worked hard to improve their technology and methods, they acknowledged the inherent limitations.

Even before the *Taney* returned, Maury explored alternatives to Walsh's sounding arrangement. If wire did not work and hemp remained unwieldy, perhaps a better option would be small, light line such as commercial packing twine, weighted by a heavy sinker. Maury arranged to have Lieutenant William Rodgers Taylor of the USS *John Adams* test this idea. By the late spring of 1850, Maury had convinced naval authorities of the importance of deep-sea sounding and the practicability of the light line method. In May the chief of the Bureau of Ordnance and Hydrography, Charles Morris, issued a circular with deep-sounding instructions for navy vessels. Under this new plan, every U.S. naval vessel would carry two sizes of twine, leads, and thirty-two-pound shot, different combinations of which would be used depending on whether or not depths or other circumstances appeared favorable for recovering the lead.[18]

Later, after 1854, Maury issued deep-sea sounding equipment only to officers who had volunteered, rather than equipping every vessel as before. Nevertheless, he maintained a vision of deep-ocean research as a democratic pursuit. To make deep-sea measurement accessible to all naval officers, he restricted acceptable sounding gear to device made of simple, inexpensive components. He never intended to rely exclusively on specially equipped research vessels to study the sea.[19]

As with his famous wind and current charts, the success of Maury's program depended on voluntary participation, in this case by naval officers commanding routine missions. A handful attempted deep-sea soundings, submitting one measurement or several or, in a few cases, more than a dozen. Captain Charles T. Platt of the U.S. ship *Albany,* one of the first vessels equipped with the sounding twine and sinkers, "entered heartily into the spirit of deep sea soundings." He reported seventy-one measurements in the Gulf of Mexico and Caribbean seas between December 1850 and February 1852. Platt's results alerted Maury to the inconsistent quality and strength of the common commercial wrapping twine he had supplied. The *Albany's* first lieutenant, William Rodgers Taylor, supervised all the deep-sea soundings, recording detailed circumstances about each experiment in his journal. Maury reprinted most of Taylor's sounding entries in order to "let others have the benefit of his experience."[20] Recognizing the need to make available to other deep-ocean investigators the raw information about numerous sounding experiments, Maury promoted deep-sea hydrography by compiling data and making it available within the maritime community, much as he had done for wind and current observations.

When he began systematic investigation of the depths, Maury strove for what he felt would be adequate, not unnecessarily numerous, data points. He discouraged sounding attempts in areas where depth measurements existed, and he published tables listing the latitude and longitude of all soundings to date, to help prevent duplicate efforts. He instructed officers not to repeat soundings within 250 or 300 miles of each other when sounding in more than 1,000 or 2,000 fathoms. He asserted that one sounding for every 75,000 square miles of ocean would enable him to make a chart showing the 1,000-fathom contour lines. This confidence that one sounding represented reliable knowledge about so much of the sea floor was called into question dramatically several years later, when several soundings near the *Taney*'s famous 5,700-fathom measurement indicated much shallower water. Soon after, Maury concluded that the *Taney*'s sounding had been inaccurate, blaming the inelastic steel wire.[21]

The year following the disappointing *Taney* voyage, Maury received permission to send out another vessel to further his research on winds and currents. In October 1851, the U.S. brig *Dolphin*, with Lieutenant Samuel Phillips Lee commanding, set sail under the same orders as the *Taney* had been given, although the sounding wire was replaced with twine. Among his other responsibilities, Lee was instructed to conduct soundings every 200 miles, both as the *Dolphin* crossed the Atlantic and on the return trip. Lee's voyage resulted in thirty-two deep-sea measurements, made between November and February of 1852. His deepest measurement was only 3,825 fathoms, much shallower than the deepest soundings made up to that date.[22]

Unlike the *Taney*, the *Dolphin* fulfilled Maury's expectations of spending most of its time at sea. Lee reported only forty days in port in eight months, "an arduous and confining cruise" by contemporary standards. The first captain to experience the grueling nature of continual deep-sea work, Lee warned that this "important service" required "patient, attentive, and laborious observations." He took special precautions to ensure the health of his crew, both because they spent so little time ashore compared to crews on other naval missions and because their work was particularly wet and strenuous. In an effort to maintain discipline, he read the articles of war each week to remind the seamen of their duties and responsibilities. Several men nevertheless deserted the *Dolphin* during the infrequent port stops, a problem that would continue to plague deep-sea research vessels.[23]

Each deep-sounding effort during these pioneering years, from 1849 until after 1855 at least, was described as an "experiment" by Maury and the officers carrying out the work at sea. Their instructions stressed the necessity to note every circumstance associated with each operation. In addition to the usual notations of latitude, longitude, time, drift, and weather conditions, hydrographers should record every instance of loss, noting the type of line used and the weight and shape of sinkers expended. Every attempt, whether or not bottom was reached, was to be noted.[24] In this way, hydrographers could compare their own soundings to others on record to determine the likelihood that they had reached bottom. They could also consult other experiments to determine which equipment and methods worked best.

The sounding method devised by Lee consisted of practices that helped prevent the sounding twine from breaking, including doubling the first several hundred fathoms of line and keeping a gentle strain on the line until it was a hundred or so fathoms under the water. He also sought more accurate results by sounding from a boat rather than the ship, a procedure that allowed him to use oars to keep the line "straight up and down." On the basis of his experience, Lee confidently asserted the practicability of sounding great depths "in any weather in which [a boat] can be lowered safely."[25] Maury concurred, and continued to encourage deep-sea sounding experiments after Lee returned to port.

Lee's soundings, as well as those made by other officers who adopted his method, rarely found such deep water as some of the early, singular efforts that resulted in depth measurements in the 5,000- to 8,000-fathom range. The majority of deep oceanic soundings in 1852 and 1853 revealed depths in the 2000- to 3000-fathom range. A bathymetrical chart Maury published in 1853 excluded several of the extremely deep soundings, such as the 5,700-fathom one from the *Taney* as well as an 8,300-fathom result from the U.S. frigate *Congress* in 1852, a 7,700-fathom sounding from HMS *Herald* in 1852, and a 6,600-fathom measurement from the *Dolphin* in 1853. Within four years, Maury had shifted decisively from celebrating to dismissing such deep measurements.

Maury's eventual decision to leave the deepest soundings off the chart was based, in part, on analysis of the materials and methods of each individual sounding.[26] Consistent overestimation of the ocean's depth arose from uncertainty over when the sinker had reached bottom during a sounding attempt. Early deep-sea hydrographers supposed that, when the lead reached the bottom, either the leadsman would feel a shock or the line would slacken and cease to run out, both of which occurred when sounding navigational depths. Maury studied the reports of the enormous depth measurements reported by the *Taney, Congress, Herald,* and *Dolphin,* as well as those that found more moderate depths. He concluded that the extremely deep ones were inaccurate because the line, rather than ceasing to run out when the sinker hit, continued to be pulled off the reel. He ascribed this effect to strong, deep undercurrents.[27] Even the moderately deep measurements that seemed reliable, Maury realized, were probably a little too large because of the difficulty of ascertaining when the sinker struck bottom.

To solve the problem of determining the moment of contact, Maury applied the method that had served him so well in developing wind, current, and whale charts. He compiled all available information about each sounding experiment and compared the time that elapsed as each hundred, or several hundred, fathoms of line ran out. From these data, he tried to find "the law of descent" that governed the behavior of a thirty-two-pound shot attached to a specific diameter of sounding twine as it moved through seawater. Constructing tables that he and other hydrographers could use to analyze individual sounding events, he demonstrated that some of the early very deep soundings simply had to be false measurements. For the first one or two thousand fathoms, these soundings had intervals that matched mean values from the compiled soundings. After that point, they diverged from the norms, indicating that the sinker had reached bottom, but the line had continued to run out more slowly than during free fall. In 1853 Maury was willing only to describe any deep-sea sounding experiment as "an approximation . . . a step in the work of measuring the depth of the ocean, by assuring us that its depth is not beyond a certain extent."[28]

BEFORE MAURY began to doubt the enormous early measures of ocean depth, he marveled at the implications for physical geography of the unexpected revelation that the ocean's depths exceeded the elevation of land. Before the *Taney*'s first, spectacular sounding, the vertical distance from the top of the Himalayas to the bottom of the sea was believed to be ten miles. With Walsh's finding, the difference was extended to twelve. The discovery of two additional vertical miles represented to Maury new territory to be explored, understood, and claimed. He insisted that the study of this newly discovered region was "certainly a matter of inquiry as profitable, as instructive and as useful as is the delineation on our maps of mountain ranges and other configurations of the earth's surface."[29]

The commercial profit to be gained from studying "that interesting country—the Atlantic slopes"—derived from the potential for submarine telegraphy. But Maury's interest in deep-sea sounding predated his involvement with the Atlantic cable project. In addition to making navigation safer by removing vigias, Maury intended to contribute to a new science, the physical geography of the sea. The deep-sea basin, he explained, represented an important focus of investigation for understanding tides, waves, and other phenomena influenced by the shape of the ocean floor. Physical geography provided the strongest intellectual motive for his experimental deep-sea soundings. Maury had in mind Alexander von Humboldt's scientific program as he arranged for naval vessels on routine patrolling missions to carry deep-sea sounding gear.[30]

The first visual representations of the sea floor, published in 1853, make clear that physical geography inspired Maury's deep-sea research program. By then, Maury felt he had enough data to construct a bathymetrical chart, which was the first ocean basin submarine contour map ever produced. He also constructed a vertical section showing the elevation of the earth from the Rocky Mountains to the Atlantic, as well as the depression below the sea level across the ocean through the Azores to Europe. The data upon which these first visual displays were based came from the soundings Maury had heralded as revolutionary for the field of physical geography. The vertical section was inspired directly by geographers' visual practice of showing a land elevation profile, a precedent that Maury acknowledged. To this, Maury appended a conjectural section of the seabed, based on the few available deep soundings.[31]

Before Maury's deep-sea sounding program, he and others had supposed that the sea floor was "regular in outline, passing gradually from deep to shallow, and from shallow to deep." Although he knew that areas near land had irregular features, he was amazed to begin receiving reports of soundings that were over a thousand fathoms different from previous ones just a few miles away. Quite suddenly, the image of the ocean floor shifted dramatically, so that Maury instructed, "we should not now be surprised if . . . observations should shew [sic] that . . . the bottom of the sea is quite irregular in its outlines, in its elevations and depressions, in its mountains and its valleys, as is the face of our continents."[32]

Upon further reflection on the erosive forces on land, felt not at all or only feebly in the depths, Maury concluded that the sea floor must be even more rugged and abrupt than dry land, subject as land was to wind and rain. Maury published another bottom profile in his 1854 volume of *Sailing Directions* that depicted a wildly irregular sea floor. The accompanying descriptions emphasized its roughness compared to the relative smoothness of land. The following year, Maury, borrowing from Clarence's dream speech in *Richard III,* painted a lurid portrait of the sea floor: "Could the waters of the Atlantic be drawn off, so as to expose to view this great sea-gash, . . . from the Arctic to the Antarctic, it would present a scene the most rugged, grand, and imposing. The very ribs of the solid earth, with the foundations of the sea, would be brought to light, and we should have presented to us at one view, in the empty cradle of the ocean, 'a thousand fearful wrecks,' with that dreadful array of dead men's skulls, great anchors, heaps of pearl and inestimable stones, . . . making it hideous with sights of ugly death."[33] The deep ocean seemed, as much as ever before, a place untouchable not only by humans, but by environmental forces as well.

As Maury developed this dramatic image of the sea floor, his hydrographers were improving deep-sounding techniques. In 1850, when deep-sea

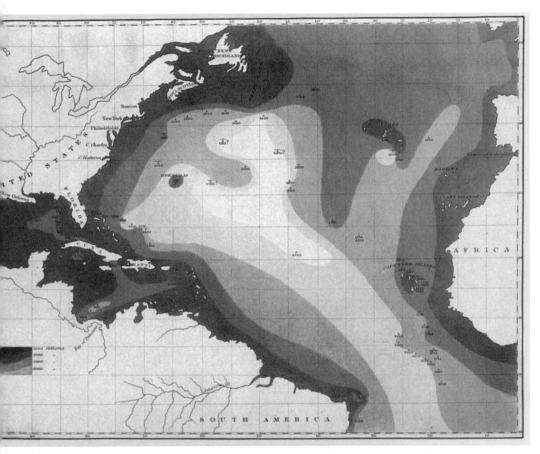

The first bathymetric chart of the Atlantic basin, 1853, based on a few dozen deep-sea soundings. (From Matthew F. Maury, *Sailing Directions*, 5th ed. [Washington, D.C.: C. Alexander, Printer, 1853], plate XIV; courtesy of the National Oceanic and Atmospheric Administration, U.S. Department of Commerce.)

sounding represented occasional, scattered experiments whose success was hardly guaranteed, Lee's orders had instructed him to "take up your position for deep-sea soundings in the calm regions known as the 'horse latitudes.'"[34] In short, Maury thought it expedient to seek out the least windy and wavy conditions possible and selected the *Dolphin*'s route using his own wind and current charts to locate precisely where the vessel would encounter the least wind and the calmest sea during the months of the cruise. When Lee returned in 1852 with a serious eye injury, Lieutenant Otway A. Berryman took over command of the *Dolphin*. His orders, though similar to Lee's, varied in one important way. For Berryman's 1852–53 voyage, Maury planned a route with something in mind other than the hydrographer's convenience. Although most of Berryman's soundings were made farther to the south, he conducted more than a dozen soundings north of a line about 45° latitude, running from the Grand Banks south of Newfoundland to the Bay of Biscay between France and Spain.

Maury's stated purpose for ordering these soundings was the navigational—and thus commercial—benefit of studying the seabed underneath the major shipping routes between the United States and Europe. The new steam vessels favored a more direct route between the continents than sailing vessels had long used, a route that lay to the north of the belts of trade winds. Berryman's soundings fell near (though to the south of) a quite similar route proposed by Atlantic submarine telegraph promoters. They hoped to lay a cable from Ireland to Newfoundland, close to the great circle, or the shortest distance across the Atlantic. No evidence exists that Maury planned this route in response to requests from telegraph promoters or naval officials. Maury's early correspondence on the subject with Cyrus W. Field, an American entrepreneur who consulted him about the feasibility of laying a trans-Atlantic cable, stated that Berryman's soundings were made to support ocean shipping. "While thus employed," Maury explained, the *Dolphin* "obtained all the information concerning the bottom of the deep sea . . . that a submarine Telegraph Company could desire."[35] Although the balance of Maury's intellectual and commercial motives for opening the deep sea may remain unclear, the prospect of trans-Atlantic submarine telegraphy immediately accelerated efforts to study the ocean floor.

GIVEN THE EMERGING image of the deep-sea floor as a forbidding place, it seems curious that plans for transoceanic telegraph cables were moving forward. Submarine telegraphy began with short sections of cable running under rivers and bays, starting in the 1840s. The 1851 Dover-to-Calais connection fueled confidence in a Mediterranean cable, finally laid in 1858. Although the idea of an Atlantic cable had been voiced before, the project awaited its champion, the American entrepreneur Cyrus Field. Submarine telegraph engineers shared hydrographers' visions of the depths as a violent and daunting environment for cables. They designed early underwater cables as stout as the lines that held seventy-four-gun men-of-war at anchor. By the time that the conducting wire had been coated with gutta percha for insulation and encased in wire armor, it was "larger than a man's arm."[36]

By 1854, though, Maury had uncovered a marvelous exception to the

A section of the Atlantic, north to south, illustrating that the sea bottom to the south of the Telegraph Plateau was too rugged to lay a cable along a proposed southern route in which one end would land in the United States. This section shows how rugged hydrographers still believed the Atlantic basin was, and by contrast how marvelous they found the exceptionally flat, moderately deep site for the proposed Atlantic cable. (From *Harper's Weekly*, Aug. 28, 1858, p. 548; courtesy of Archives & Special Collections at the Thomas J. Dodd Research Center, University of Connecticut Libraries.)

rugged, dangerous surface of the sea floor. That year, he sent a special report to the Secretary of the Navy addressing "the practicability of a submarine telegraph between the two continents *in so far as the bottom of the deep sea is concerned.*" Maury's report coincided with a request from Field for his opinion on the subject.[37] Two discoveries led to Maury's sudden, new conception of the sea floor as an environment ideally suited for a telegraph cable. One was the fortuitous discovery made by Berryman, in the summer of 1853, of relatively moderate depths along the great circle shipping route. Since Columbus, navigators leaving Europe had sailed south to Madeira, the Canary Islands, or the Azores and then traveled west, taking advantage of the trade winds. The advent of steamers drew shipping traffic north toward the great circle route, the shortest distance between Europe and North America. For the same reason, submarine telegraph promoters favored this route for the Atlantic cable, which would stretch 1,600 miles from Valentia, Ireland, to St. John's, Newfoundland. The second discovery, one that strengthened Maury's confidence in trans-Atlantic telegraphy, was Berryman's return with the first series of bottom samples from the great depths.

According to Maury's gleeful report, Berryman's line of soundings decisively demonstrated that the sea floor between Newfoundland and Ireland, the two proposed landing areas, "is a plateau which seems to have been placed there especially for the purpose of holding the wires of the submarine telegraph, and keeping them out of harm's way."[38] By this, he meant that in contrast to other parts of the Atlantic, which were still believed to measure incredible depths of 5,700 fathoms and more, the proposed route was only moderately deep, in the 2,000-fathom range. Thus less cable would be required than would be necessary to cross some of the deeper parts. Yet the cable would rest safely, out of the reach of anchors and fishermen's trawls, in what was, after all, very deep water.

During the time Maury had been gripped with the excitement of constant new discoveries about the ocean floor, but in search of verification of the soundings, John M. Brooke, who subsequently accompanied the North Pacific Exploring Expedition, had suggested his simple but effective device to recover bottom sediment. Brooke had served in the Coast Survey in 1849 under Samuel Lee, before Lee commanded the first *Dolphin* voyage. At that time, Brooke learned how to obtain bottom sediment, one of the regular surveying practices in the Coast Survey since the 1840s. These samples served navigational purposes and generally came from depths no greater than 100 fathoms. Although Lee had tried to retain Brooke for an additional year on the Survey, Brooke was transferred to a ship operating around Africa until 1851, when he was ordered to the Naval Observatory. There, in 1853, he proposed to Maury that the thirty-two-pound shot sent down as sinkers for deep-sea sounding lines be rigged to detach at the bottom. That way, the smaller and lighter sounding lead, with its bottom sample, could be recovered without breaking the line.[39]

Brooke's first idea involved attaching a knife to a moveable rod, which passed through the shot, so that the knife sliced the lines cradling the shot. Thus the shot was left behind when the lead and sounding line were hauled in. The folly of using open knives near sounding lines rendered that design impracticable, but Brooke's next version proved workable. The new device utilized the same basic arrangement as before, namely a shot with a hole bored or cast through the center, into which a metal rod was inserted. The shot rested on a sling held at the top by a metal device like ice tongs, which opened when the sounding line slackened, after the shot struck bottom. Sea

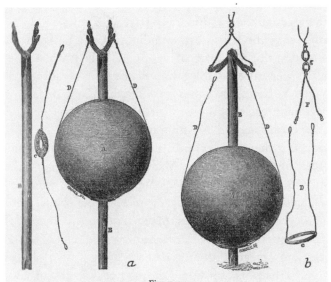

Figure 4

[Brooke's Deep-sea Sounding Apparatus for bringing up specimens of the bottom: a, ready for sounding; b, at moment of release on reaching bottom.]

John M. Brooke's deep-sea sounding device, which revolutionized ocean measurement by permitting confirmation that the sounder had reached bottom. When the rod struck the bottom, the detaching mechanism released the wire cradle holding the sinker so that only the rod and a small bottom sample returned to the surface. (From Matthew F. Maury, *Physical Geography of the Sea* [New York: Harper, 1855], 287.)

trials, including Berryman's from the *Dolphin* as well as Brooke's own during the North Pacific expedition, proved the sounder's effectiveness. It combined the simplicity and the economy that Maury insisted upon for deep-sea sounding gear.[40]

Brooke's device revolutionized oceanic sounding by enabling deep-sea soundings to be confirmed for the first time ever. The armed lead, freed of the weight of the sinker, could be recovered to observe the presence or absence of bottom sediment, proving whether or not the lead had struck bottom. Maury's first bathymetric chart relied on data from Brooke's sounder, especially to rule out the use of several early deep measurements. The chart included shaded zones marking contours of one, two, three, and four or more thousand fathoms. Admitting that the shape of these curves was, "for the most part, a matter of conjecture," Maury nonetheless wished to convey "a general idea as to the shape of the Atlantic basin." Most of all, he hoped that the chart would "excite an interest among officers" and inspire them to conduct more deep-sea soundings. In the following years, more detailed bathymetrical charts appeared. Instead of 90 soundings, of which almost half were made in the Gulf of Mexico and Caribbean seas, the 1855 chart recorded about 189 soundings, or three times the number of measurements for the Atlantic basin as the original chart displayed.[41]

Results from Brooke's sounder confirmed Maury's initial judgment that the ocean floor at the great circle was uniquely suited for safely holding telegraph wires. Berryman's northernmost soundings, about nineteen of them, suggested to Maury the presence of a relatively shallow steppe, which he optimistically christened "Telegraph Plateau." The ability of Brooke's sounder to retrieve bottom samples proved crucial. The surface of the plateau, Maury was delighted to discover, had "a down-like softness."[42] Although the officers who conducted the soundings had judged the sediment to be merely clay, they nevertheless preserved, labeled, and returned samples from each sounding to the Depot. Maury sent specimens to the microscopists Christian G. Ehrenberg and Jacob Whitman Bailey, who discovered that the sediments consisted entirely of microscopic shells, with not a particle of sand or gravel. The samples, Maury concluded, "teach us that the quiet of the grave reigns everywhere in the profound depths of the ocean; that the repose there is beyond the reach of the wind; it is so perfect that none of the powers of the earth, save only the earthquake and volcano, can disturb it."[43]

This quiescent image of the deep-sea floor diverged dramatically from the previous wild, rugged one. Interest in physical geography, with attention to the forces of erosion and analogies to land geography, had shaped the first few, occasional deep-sea observations into a picture of the ocean bottom as a forbidding landscape. Although for Maury, as for other promoters of science, knowledge about the physical world was enough of an end to justify deep-sea study, this rather abstract reason gave way to demonstrably practical motives with satisfying swiftness. Within just a few years of the *Taney's* first oceanic sounding, Maury could ask rhetorically of deep-sea investigation, "What is the use of a new-born babe?" only to answer that his results "forthwith assumed a practical bearing . . . with regard to the question of a submarine telegraph across the Atlantic."[44]

The needs of submarine telegraphy guided the reinterpretation of the shape of the ocean floor that accompanied the new measurements made by Brooke's sounder. Hydrographers and submarine telegraph entrepreneurs accepted the benign image of the "Telegraph Plateau" with alacrity. A report published by the Atlantic Telegraph Company to promote the first cable-laying effort in 1857 made the bold claim that "Nature, indeed, has made every necessary preparation for the work."[45]

By the mid 1850s, entrepreneurs and investors were demanding knowledge of the sea floor before embarking on such a fanciful venture as

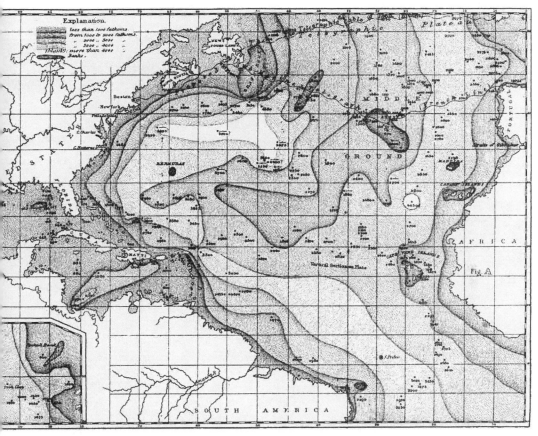

A bathymetric chart of the Atlantic basin incorporating measurements made with Brooke's sounder, including those made along the Great Circle route proposed by submarine cable entrepreneurs. This chart displays three times more soundings than Maury's first bathymetric chart. (From Matthew F. Maury, *Sailing Directions,* 7th ed. [Washington, D.C.: William H. Harris, Printer, 1855], plate XV; courtesy of the National Oceanic and Atmospheric Administration, U.S. Department of Commerce.)

stringing a wire between the Old and New Worlds. The U.S. and British governments both required reassurances before promising subsidies, the loan of ships, and other resources to help lay the cable. The process of convincing potential participants of the practicability of a trans-Atlantic cable involved broad public education about the deep sea, which encouraged people to visualize a telegraph cable lying safely in a place they had never previously imagined.

Submarine telegraph promoters faced many technical challenges and unanswered questions. Cyrus Field, who promoted the idea of a trans-Atlan-

THE GREAT WORK OF THE AGE.

Telegraphic Union of the Old and New World.

THE FIRST CABLE EVER CONSTRUCTED.

Arrangements for Laying the Last Cable.

PROGRESS OF SUBMARINE TELEGRAPHING.

PICTURES OF THE CABLES OF THE WORLD.

The Telegraph Plateau of the Atlantic.

THE INFUSORIA OF THE PLATEAU.

Microscopic Specimens from the Bed of the Ocean.

The Effects of the Success of the Submarine Telegraph.

The Daily Morning and Evening Publication of European News in the New York Herald.

THE DEPARTURE OF THE NIAGARA.

THE FIRST GULF CABLE.

THE SECOND GULF CABLE.

THE GREAT ATLANTIC CABLE.

TOTAL LENGTH OF SUBMARINE CABLES ALREADY LAID DOWN.

INFUSORIA OF THE TELEGRAPH PLATEAU.

EFFECTS OF THE SUCCESS OF THE

THE TELEGRAPH PLATEAU OF THE ATLANTIC.

MORE SUBMARINE CABLES PROPOSED.

tic cable steadily for a dozen years, questioned electricians, engineers, and scientists before deciding to invest his money and energy. Along with the critical concern about whether an electrical signal could survive transmission along so long a wire, Field investigated the question of the safety of a cable on the ocean floor. At the outset, he knew nothing of either the conditions at great depths or the geography of the intended route. In February 1854, he turned to the only person in the world who had published descriptions of the Atlantic bottom. To Field's query Maury responded confidently that the Atlantic telegraph could go ahead without reservation. He described the ocean along the proposed route as deep enough to keep the cable safe, but not so deep as to require an impossibly long cable or to make the task of laying it overly difficult. In addition, Maury attested to the quiescence of the sea floor and the nonabrasive nature of the bottom sediments. Until 1858 Maury frequently advised Field about the results of sounding voyages and microscopic examinations of bottom sediments.[46]

Field relied on Maury's testimony to calculate the depth and type of terrain over which engineers would install the cable. Cable manufacturers needed to know how many miles of cable to construct, as well as where along the route to use armored cables and where thinner deep-sea cables would suffice. Maury also advised the Atlantic Telegraph Company about the best route and the most propitious time of year for laying the cables to avoid dangerous North Atlantic storms and fogs.[47] Beyond these technical specifications, Field used Maury's testimony to convince shareholders, potential investors, and government officials that the ocean floor would offer a safe home for a cable.

Visual representations and bottom specimens were employed to help cable entrepreneurs construct an inviting image of the deep sea. Newspaper articles from before and during cable-laying attempts in the 1850s showed readers a picture of the ocean in which they could imagine a cable. Newspapers devoted whole front pages to cable-laying voyages during 1856 and 1857. These featured images depicting the bottom as safe, still, and free of dangers. Bottom profiles were common and drawings of microscopic views of bottom sediments with appropriately reassuring

The Atlantic cable-laying plans and voyages, whether successful or not, were front-page news. The microscopic views of Atlantic bottom sediment from the Telegraph Plateau also featured prominently in promotional literature for the project. (From the *New York Herald*, April 26, 1857; courtesy of the General Research Division, New York Public Library, Astor, Lenox, and Tilden Foundations.)

prose were prominently displayed: "These ingravings [sic] represent the infusoria magnified three hundred times their natural size and are so infinitesimal as to be the merest mites on the surface of a microscopic glass. Notwithstanding they are so perfect in form, delicate in construction, and so minute in size, the bed of the plateau is so quiet and undisturbed from the action of the ocean, that scarcely any of them, comparatively speaking, are injured or broken by abrasion or attrition. They will, indeed, form a sort of bed of down for the cable to rest upon." Lest readers fear exaggeration, journalists emphasized, "incredible as this may appear, it is nevertheless true in every particular."[48]

Actual pieces of cable and deep-sea bottom samples acted as tangible testimony to the project's viability. At a meeting in November of 1856, underwriters expressed serious concerns about the safety of the cable. In answer to their questions, Cyrus Field produced specimens of the minute shells that the sounding voyages had recovered and passed them around, explaining that these nonabrasive particles covered the bottom. "Upon this bed of small shells the cable . . . would rest forever," he assured them. Also that year Field attended a Royal Geographical Society meeting, bringing along not only a profile of the bottom, but also a portion of the cable and specimens of Atlantic deep-sea mud.[49] Taken together, these artifacts made it easier to imagine a cable resting safely and peacefully in a previously unimaginable place.

In the 1850s, promotional literature for the Atlantic cable addressed the question of danger on the sea floor and, both visually and verbally, presented an attractively benign picture of the depths. The frontispiece of an 1857 publication by the directors of the Atlantic Telegraph Company showed a north-south section of the Atlantic, prominently labeling the quite obvious "Telegraph Plateau of Lieut. Maury," and an east-west section of the plateau itself, along the proposed cable route. A second east-west section showed the topography 1,500 miles south along a route advocated by those who wanted one end of the cable in American territory. Predictably, this alternate route was full of jagged peaks and abrupt rises, emphasizing the text's claim that the great circle route was preferable. This pamphlet also emphasized the safety and softnesss of the plateau. Examined under "powerful microscopes," the material that covered the "Atlantic steppe" was found to be skeletons of "passed away generations of living beings." Cable companies could comfortably dismiss the question of whether life existed at great depths by invoking the then-current azoic theory of Edward Forbes. The pamphlet concluded that "if art had prepared a bed for an oceanic cable, after full deliberation, it could not have devised

The famous, widely reproduced depiction of Matthew Fontaine Maury's Telegraph Plateau, an east-west section across the Atlantic showing the *Niagara* and *Agamemnon* laying the cable from the center toward the two continents. (From *Harper's Weekly,* Aug. 28, 1858, p. 548; courtesy of Archives & Special Collections at the Thomas J. Dodd Research Center, University of Connecticut Libraries.)

any more complete arrangement than this profound recess of still water, paved beneath with smooth, impalpable powder."[50]

Another book, purported to have been written by the 1858 Atlantic cable itself—"Mine has been a short but most eventful career . . ."—described its "ocean bed" as a gently leveled submarine plain that had been "very accurately examined by sounding," whose "surface is devoid of all abrupt irregularities." Invoking the same powerful microscopes, this volume also assured readers that the bottom was safe and nonabrasive, composed of "one-celled elementary organisms which afford the battle-ground of learned philosophers, who are striving to settle the boundary question of the vegetable and animal domains."[51] The implication was clear: Any place where philosophers battle could not endanger a submarine cable.

The most famous image of the ocean floor was Maury's "Telegraph Plateau." In his 1854 letter to the Secretary of the Navy, Maury described Berryman's line of soundings in the *Dolphin* the summer before as *"decisive"* about the prospect of laying a submarine cable across the Atlantic. In a depth "neither too deep nor too shallow," the wire would be permanently out of danger from anchors and icebergs and yet easy to lay. The bottom he characterized as "quiet . . . as it is at the bottom of a millpond." He quoted Bailey's report, noting that "the eminent microscopist" had been surprised to learn that the soundings were full of microscopic shells, with "not a particle of sand or gravel" in them. In his popular book, *The Physical Geography of the Sea,* Maury spun a life-cycle analogy to describe the ocean floor as the silent graveyard for all the ocean's inhabitants: "Henceforward, we should view the surface of the sea as a nursery teeming with nascent organisms, its depths as the cemetery for families of living creatures that outnumber the sands on the sea-shore for multitude."[52]

People involved in both the 1857 cable-laying attempt and the 1858 success acknowledged and appreciated Maury's contributions. At a dinner celebrating the arrival of the first message across the Atlantic cable, Field credited Maury's part in the endeavor's success by saying, "I am a man of few words: Maury furnished the brains, England gave the money, and I did the work."[53] That cable failed a month later, however, and Maury would get no such acknowledgment in 1866. He spent the years in which he might have focused once again on probing the deep ocean instead developing underwater torpedoes and otherwise supporting the Confederacy.

THERE IS ANOTHER reason, however, that Maury's name did not have a prominent place in 1866 acknowledgments. In the 1860s, cable entrepreneurs, advocates, and problem solvers ceased to worry about the deep sea, which they had judged an inert, secure place for telegraph wires. Shareholders' meetings no longer devoted time to questions about the sea bottom. Nor did newspapers, scientific or technical journals, or books about the cable venture expend any further effort defining the ocean floor, either verbally or graphically. In 1859 and 1860, the British government and the Atlantic Telegraph Company held a joint inquiry into the construction of the submarine telegraph. Most testimony concerned electrical problems posed by a 2,000-mile-long wire, the mechanical challenges of laying thin cables in deep water, or the manufacturing concerns about producing long wires of consistent quality. In nine months of hearings, the nature of the ocean floor appeared in the testimony of only nine men in seven sessions, and four other witnesses briefly addressed the issue of whether deep-sea life endangered cables. Although the committee affirmed the importance of sea-floor surveys to guard against potential mechanical or chemical injury to cables, the report found the general level of knowledge about the sea floor adequate for cable-manufacturing and cable-laying standards.[54]

Indeed, though promotion of knowledge about the depths had been intense before the 1857 and 1858 cable-laying attempts, it was almost absent afterward. The first flush of confidence following the 1858 success convinced the cable industry that the ocean represented no insurmountable barrier to telegraph wires. "The great fact that the cable is laid without accident, and that its insulation and transmitting power are reported perfect, render the question of plateau or no plateau . . . of secondary importance to the public."[55]

When George Wallich sailed on an 1860 cable survey, his warnings that the expedition was not learning enough about the ocean floor went entirely unheeded. His ship, HMS *Bulldog,* sounded an alternate cable route, from the Faroe Islands to Greenland to Iceland. Its advocates lacked confidence in the ability of electrical engineers to make a long cable function well. They therefore preferred a route whose several underwater portions were each much shorter than the 2,000-mile stretch across the Atlantic. Neither promoters of the alternate route nor the Admiralty harbored concern about the nature of the sea floor beyond its contours, so they did not plan to take a scientist on board. Wallich sailed only because he lobbied energetically to accompany the expedition. He complained that Captain Leopold McClintock often used sounding devices and methods that did not allow for recovery of bottom sediments. McClintock, not swayed, declined to order more soundings with sampling devices. Nor were promoters ashore worried about the composition of the bottom. When the *Bulldog* results were compiled and published, bottom profiles occupied center stage, and almost no notice was taken of the nature of bottom sediments.[56]

When the Atlantic Telegraph Company regrouped for another attempt at the Ireland to Newfoundland great circle route, it stressed that when the first attempt had been made in 1857, "little or nothing was known of the bottom of the Atlantic." Following the 1858 failure, the company identified the several miles closest to each shore as the relatively unknown and worrisome places, and reassured shareholders that these areas had since been scrutinized by two British government expeditions. The deep-sea floor, by contrast, was no longer problematic. In May 1862, Cyrus Field discussed the status of the cable project with the American Geographical and Statistical Society. Although he did bring a specimen of the new cable to explain the improvements from the 1858 version, he dismissed any doubts about the ocean floor by asserting that "neither length of distance nor depth of water are any insuperable impediments." In 1863, a company called Universal Telegraphic Enterprise advocated the southern route for the Atlantic cable, between Cape St. Vincent and Cape San Roque, so that one end of the cable would land in American territory. The company's prospectus contained only one sentence about the ocean's depths, claiming an absence of currents at the bottom.[57] In terms of cable laying then, the ocean floor ceased to be a critical problem by the early 1860s. It was shelved as solved, replaced by more pressing electrical and mechanical problems. This did not, of course, end hydrographic interest in the deep sea, or the desire of cable companies to promote government hydrographic surveys of other prospective cable routes.

As ENTREPRENEURS pursued the second set of attempts to lay the Atlantic cable in 1865 and 1866, Britain took over leadership of deep-sea exploration. Like the hydrographers working under Maury, their British counterparts extended knowledge of the deep-sea floor and developed better technological means to reach the ocean's depths. In 1854, Lieutenant Samuel Phillips Lee had published his book-length report, funded by the U.S. government, on the Atlantic *Dolphin* voyage. Read and cited by British as well as American hydrographers, this report became the first of many such official publications that passed along experience and innovations in deep-sounding technology.[58] Initial U.S. dominance of deep-sea sounding rankled, as did all American scientific and technical successes that impinged on British mastery of the sea. With the end of the Crimean War in 1856, and especially after the outbreak of the Civil War in the United States five years later, the British Hydrographic Office took on the mantle of patron for those attempting to conquer the depths by expanding its deep-sea surveying, especially in the Mediterranean and North Atlantic, areas targeted by submarine telegraph companies.

The Hydrographic Office had assumed a scientific character under the leadership of Sir Francis Beaufort, who served as Hydrographer of the Navy from 1829 to 1855. His predecessor, the famous Arctic explorer Rear Admiral Sir William E. Parry, had organized pendulum experiments and magnetic observations of great interest to men of science. Parry's explorations benefited the whalers and sealers who worked in the Arctic more than they did the merchant fleet. In contrast, Beaufort concentrated on improving charts and sailing directions for regular shipping and passenger routes. Like Maury, Beaufort combined scientific interest in meteorology and ocean currents with a practical commitment to promoting safer navigation. Thus while his accomplishments included devising a scale for measuring wind force by eye and standardizing chart symbols, he also facilitated scientific research in natural history, geology, meteorology, and other fields, work that could be accomplished alongside hydrographic surveying. He helped, for example, convince the Lords of the Admiralty to publish *The Admiralty Manual of Scientific Inquiry* in 1849. He also encouraged naval surgeons such as Thomas Henry Huxley to pursue independent scientific research.[59]

Beaufort's successor as Hydrographer, Rear Admiral John Washington, directed British attention below the waves. Soon after taking office in 1855, Washington became aware of Maury's bathymetrical chart. Like Maury, he used his position to extend knowledge about the sea floor in anticipation of the cable. He ordered Lieutenant Joseph Dayman to conduct surveys between Ireland and Newfoundland in the *Cyclops* in 1857 and the *Gorgon* in 1858. In the wake of the failure of the 1858 cable, he sent Captain McClintock in the *Bulldog* to explore an alternate route, via the Faroe Islands, Iceland, Greenland, and Labrador. Finally, after several years' hiatus, Washington ordered an additional cable survey in 1862, when Captain Richard Hoskyn took HMS *Porcupine* to examine the troublesome first 300 miles of the great circle cable route from Ireland. Cable engineers hoped that he would find a more gentle slope than the sharp drop from 550 to 1,750 fathoms encountered during the previous laying attempt. These few voyages were only the beginning of an efflorescence of British exploration of the deep sea, increasingly the object not only of hydrographic surveyors but of naturalists as well.

chapter four

A Sea Breeze

Hurrah for the dredge, with its iron edge
And its mystical triangle,
And its hided net with meshes set,
Odd fishes to entangle!
The ship may rove through the waves above,
Mid scenes exciting wonder;
But braver sights the dredge delights
As it roves the waters under!

Then a dredging we will go, wise boys!
Then a dredging we will go!

—Edward Forbes, "The Dredging Song," *Literary Gazette*,
November 7, 1840

NATURAL HISTORY of the ocean began modestly, at the shoreline. Beachcombing, to search for beautiful and unusual shells, gained popularity in the eighteenth century and remained a central feature of marine natural history. Naturalists in serious pursuit of answers to traditional questions about natural order, or new inquiries about geographical distribution and the relationship of contemporary animals to fossils, restlessly sought novel forms beyond the confines of the beach.

Natural-history dredging grew popular not in tropical seas, but in the cold waters around Scandinavia, on the fringes of Britain, and off the shores of New England. British naturalists inaugurated the first sustained, nationally organized investigation of marine zoology. Dredging became a favorite pastime among vacationing naturalists and a national project of the British Association for the Advancement of Science.

The pioneers of marine zoology first used dredges to collect specimens from the unglamorous decks of rowboats and fishing vessels, rarely from the glorified vantage of naval ships on exploring expeditions. Marine natural history soon after came of age aboard yachts, promoted as much by the growing social appeal of seaside holidays and ocean travel as by the intellectual appeal of marine fauna to midcentury science.

THE EIGHTEENTH century saw the rise to fashion of shell collecting, reflecting a new admiration for nature that departed from the Enlightenment desire to tame and rule nature. Collectors sought for their cabinets natural objects of great beauty, including birds, moths, flowers, and fossils, in addition to shells. Serious conchologists, starting with Linneaus, collected and studied only the shells, not the mollusks' soft bodies. Classification was not based on anatomy, or on juvenile forms, knowlege of which was largely unknown until well into the nineteenth century. Conchologists began in the 1820s to reject Linneaus's system for mollusks and adopt Lamark's system for classifying shell and animal together. French savants initiated scientific study of invertebrate anatomy. In Britain, extensive practice of preserving wet specimens waited for the 1845 repeal of excise duties on glass levied during the Napoleonic war. The tax increased the cost of glass fourfold, to the great detriment of medicine and natural history. Within a few years after the law's repeal, marine zoology became more widely accessible to a broader social spectrum of naturalists.[1]

Serious naturalists and vacationing dabblers alike initially searched for specimens along the shore. Experienced beachcombers learned to visit remote areas, particularly after storms, which threw unusual creatures from deep water up onto shore. The first dredgers tended to be those who grew up on coasts, with personal experience on small rowboats and sailboats, and social access through work or family to fishermen or other watermen. They, like many naturalists, worked alone or with a local friend, collecting specimens to study or exchange with other naturalists.

Natural-history dredging was first recorded in the late seventeenth century, by Italian naturalists who dredged in very shallow water. In the 1770s, the Danish naturalist Otto Frederic Müller adapted a common oyster dredge to collect marine invertebrates. From that time on, individual naturalists, especially from northern countries with significant fisheries, employed dredges occasionally to study the fauna of their local waters. Dredgers quickly discovered that they found rare or new species in deeper water.[2]

In the 1830s, Edinburgh emerged as a center of dredging activity. There and elsewhere, field classes were added to natural-history curricula.[3] A group of professors and students interested in marine organisms began to incorporate trips on fishing vessels into their field outings. Participants in dredging excursions included not only Professor Robert Jameson and the zoologist Robert Grant, but also their students: Edward Forbes; John Goodsir, a collector of marine invertebrates since childhood; Charles Darwin; George C. Wallich, the microscopist who later went on HMS *Bulldog*'s cable survey; and George Johnston, an active promoter of marine zoology in Northumberland who founded the Berwickshire Field Club.

Forbes would become the most energetic proponent of natural-history dredging in Britain. He arrived in Edinburgh in 1831 to undertake medical studies, but gravitated away from medicine, partaking enthusiastically in collecting jaunts. He brought to these excursions his boyhood experience gathering and studying the flora and fauna of the Isle of Man. His father's involvement with the local fishery industry no doubt accounts for his proficiency at sea. As a boy, Forbes occasionally accompanied fishing vessels to a scallop bank five miles off shore, where he dredged for invertebrates in water up to twenty fathoms deep. During his university holidays, he continued dredging from rowboats around the Isle of Man, publishing his first major scientific paper as a result. Forbes often dredged with John Goodsir, with whom he shared a flat that was as much museum and workroom as living space.[4]

For the naturalists who began investigating marine zoology, the dredge was their primary tool. Forbes described the dredge as "an instrument as valuable to a naturalist as a thermometer to a natural philosopher." The dredge also defined a new, and growing, community of naturalists, as indicated by Forbes's song, "Hurrah for the Dredge." Use of the dredge defined ocean sampling as naturalists' rather than hydrographers' work. Naturalists declared, "To afford us any certain knowledge of higher forms, recourse must be had, not to the sounding line, but to the dredge." Even microscopists, for whom sounder samples provided sufficient materials for study, recognized zoologists' need for collecting gear other than sounders. Christian G. Ehrenberg predicted, "By and by we shall learn how to bring up larger samples of the bottom, and bring up some Leviathans."[5]

Naturalists' dredges derived from those used by oyster fishermen, with the important innovation of reducing mesh size to recover even small objects and creatures. By contrast, oystermen used large mesh to allow younger and smaller animals to fall back to the sea floor alive. Forbes's instructions for using a dredge reveal his familiarity with, and dependence on, people who worked at sea: "I got an old fisherman to superintend the work . . . Have plenty of good rope (the sailors will tell you what kind) & fix the rope by a strong knot to the ring of the dredge—a rowboat & a few good rowers will complete the operation. When the dredge begins to feel heavy pull it up."[6] Forbes assumed that his readers would either recognize the heft of a full dredge or know where to hire someone who did.

British dredgers first adopted "the naturalist's dredge," the one originally devised by Otto F. Müller. Before about 1840 "naturalist's dredges" were based on whatever style dredges local fishermen used. If a particular variation was adopted by other naturalists, it was christened with that person's name. Thus George Wallich mentioned using "Seton's dredge," on an excursion with Forbes during their student days in Edinburgh. In 1838, Robert Ball, of Dublin, started experimenting with dredge designs of his

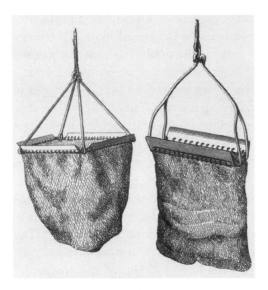

Otto Frederick Müller's dredge (left) and Robert Ball's dredge (right). Naturalists' dredges were derived from those used by oystermen, but modified to capture smaller specimens and to be lighter and more portable for holiday travel. (From Charles Wyville Thomson, *The Depths of the Sea* [London: Macmillan and Co., 1873], 239, 240.)

own. Six years later, Forbes announced that Ball's newest design, thereafter called simply "Ball's dredge," made Müller's model obsolete.[7]

Two innovations transformed the dredge from fishermen's tool to naturalists' instrument. First, the naturalists' dredge was rectangular instead of square, and it had scrapers along both long sides of the opening instead of only one side. This arrangement ensured that the dredge would operate no matter which side touched bottom first, a feature that compensated for naturalists' lack of skill. It also made dredging easier to learn. Second, the arms folded down for storage or transport, a feature that permitted the dredge to be "packed up and carried anywhere in the bottom of a carpet bag."[8] Such portability proved a definite asset for the gentlemen-dredgers who spent summer holidays traveling around northern Britain to dredge.

Like the British, Americans embraced scientific dredging. At first dredging was practiced by amateurs and beginners who grew up near the sea; only later did promoters located in central institutions adopt the practice. The most active American naturalist-dredger was William Stimpson, who sailed as naturalist with the North Pacific Exploring Expedition. He devoted his career to marine zoology, and raised marine specimens in aquaria before British experiments with keeping marine animals alive in glass tanks, carried out in the 1850s by Robert Warington. An amateur naturalist from his boyhood in Boston, Stimpson knew enough about boats and mariners to arrange a stint for himself on a Newfoundland Banks fishing vessel in order to escape a hated surveying job. Although the seamen worked the greenhorn hard, Stimpson relished the trip and, by his return, had resolved to continue his studies of marine fauna over his father's objections to a scientific career.[9]

Stimpson attended the August 1849 meeting of the American Association for the Advancement of Science (AAAS) and joined a small party led by Louis Agassiz for a post-meeting cruise on a steamer. Despite his experience, Stimpson appreciated Louis Agassiz's tips on dredging technique. Later that month, Stimpson and Charles Girard demonstrated dredge use at a field meeting of the Essex Institute in Salem, Massachusetts. Stimpson recognized the tremendous zoological value of dredging, while Agassiz, who was interested in marine zoology but hardly exclusively so, was delighted to find an energetic young person willing to undertake the labor. In October 1850, Stimpson became a student of Agassiz's. He spent the next two summers dredging off the New England and northern coasts, especially in the area of Grand Manan, Canada, an island in the entrance to the Bay of Fundy, about whose fauna he published the paper that established his scientific reputation internationally. During the winters, Louis Agassiz moved to Charleston, South Carolina. In 1852, Stimspon accompanied him and dredged for several months in southern waters. Within that year, Stimpson was selected as the naturalist for the North Pacific expedition on the basis of his extensive experience.[10]

Stimpson resembled Forbes in the fervor with which he deployed the dredge, but he lacked sufficient standing within the scientific community to promote its use in the United States the way that Forbes did in Britain. He did acquaint his other mentor, Spencer Fullerton Baird, with the benefits of dredging. As assistant secretary of the Smithsonian Institution, Baird acted as the American distributor of dredging equipment and other collecting gear for travelers and naturalists who agreed to seek out marine fauna.[11] But, contrary to the situation in Britian, dredging in the United States remained mostly a preoccupation of local natural-history organizations such as the Essex Institute and the occasional pastime of scattered naturalists.

THE GROWTH OF marine zoology owed its vigor to the social appeal of the ocean. By the 1830s, a new breed of middle-class naturalists appeared on the scientific scene who were more down to earth and earnest than the earlier wealthy cabinet collectors. There were also more of them. Collecting and studying animals and plants appealed to the prosperous middle classes as a rational, morally appropriate form of entertainment.

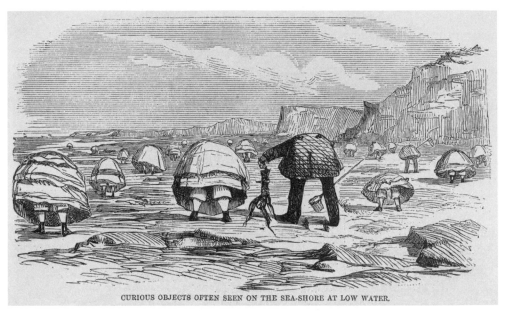

CURIOUS OBJECTS OFTEN SEEN ON THE SEA-SHORE AT LOW WATER.

A cartoon showing Victorian beachcombers at the seashore at low tide. (From *Harper's Weekly,* Sept. 11, 1858, p. 592; courtesy of Archives & Special Collections at the Thomas J. Dodd Research Center, University of Connecticut Libraries.)

"Closet" naturalists gave way to those willing to carry nets, bottles, bags, guns, and other collecting implements out into the field. As a suitably cultural and socially desirable activity, science filled the leisure time of provincial businessmen, prosperous merchants, and others aspiring to climb the social ladder.[12]

From the ranks of Victorian middle-class holiday-goers came an army of beachcombers, amateur naturalists who prowled the shorelines, inspecting and collecting shells, seaweed, and marine creatures. They represented the market for the numerous books published around midcentury on marine natural history, including Philip Henry Gosse's *The Ocean* and *A Naturalist's Rambles on the Devonshire Coast,* and Charles Kingsley's *Glaucus; or, The Wonders of the Shore,* as well as Elizabeth Cary Agassiz's *Seaside Studies in Natural-History,* which she wrote with her stepson, Alexander.[13]

Many of these books were aimed at women and children, for whom the study of marine animals was deemed moral and uplifting. Collecting marine animals and seaweeds provided wholesome activity during beach holidays. Margaret Gatty, whose contributions to algology were recognized and appreciated by leaders in the field, started studying seaweed during a stay at the coastal town of Hastings, on the Strait of Dover in southeastern England, to recover from the birth of her seventh child and a subsequent bronchial ailment. Bored by her unusual inactivity, Gatty read William Harvey's *Phycologia Britannica* and began wandering the beach to look at seaweeds for herself. She returned home to Ecclesfield, in Yorkshire, with a passion for their study and began writing popular natural-history stories, ultimately publishing the respected *British Seaweeds* in 1863. Gatty recruited her whole family for the project of collecting, turning holidays into excursions to search for unusual specimens.[14]

Although Gatty and her family pursued marine natural history more seriously than most, they were joined on the strand by many middle-class families seeking morally appropriate leisure activities. Families could also enjoy the novel hobby of keeping marine animals alive back at home with that new equipment for science and entertainment, the aquarium. By the mid 1850s, largely thanks to the extensive dredging networks that had developed, London had two suppliers of live animals as well as a public

A cartoon with the caption "Terrific Accident. Bursting of Old Mrs. Twaddle's Aqua-Vivarium. The Old Lady may be observed endeavoring to pick up her Favourite Eel with the Tongs, a work requiring some address." (From *Punch*, Dec. 19, 1857, p. 250; courtesy of Lee Jackson at http://www.victorianlondon.org.)

aquarium. Within two decades, Britain had almost a dozen public aquaria, and all major European cities had also built them.[15]

The popularity of dredging related to the growing interest in outdoor adventure. Especially around midcentury, vigorous outdoor exercise earned the endorsement of the new professional classes, from whose ranks came hobbyists, serious amateur naturalists, and the new career scientists. Field collecting resonated with contemporary commitment to recreation that strengthened both the body and the mind. It not only promoted health but also instilled in the collector an appreciation for order in nature as evidence of God's design. Along with geology, entomology, and ornithology, naturalists also praised conchology, the study of mollusks (later called malacology), on religious grounds. Charles Wyville Thomson, a dredger who became the chief scientist on the *Challenger*, declared, "A grand new field of inquiry has opened up, but its culture is terribly laborious." At their desks and laboratories, young naturalists eagerly anticipated "sea trips." The difficulties "add to the zest of the research," declared Forbes.[16]

Dredgers told stories about their adventures that provided as much instruction as entertainment. George S. Brady, a Newcastle-area naturalist and microscopist, described to Joshua Alder a dredging excursion off Galway, Ireland, in 1865. He characterized one of the days "as almost entirely a lost day as the dredge got foul of a rock shortly after commencing." He and his companions spent an hour pointlessly tugging on the dredging rope without budging the dredge. Rather than risk losing it by straining the line further, they attached an oar to the rope to keep the end afloat. After the tide turned, they returned to the dredge and managed to extricate it, much to Brady's relief.[17] Such stories not only provided tips for future dredgers, but also promoted the attitude that misadventures provided learning experiences.

The shared experiences of dredging excursions created an intense camaraderie. Central to the burgeoning identity of dredgers was the display of manly courage in the face of physical danger or discomfort. Forbes listed the ingredients for successful dredging operations: "a good boat, plenty of zeal and leisure, unusually fine weather, and a strong stomach." George Johnson wrote to his friend William Thompson about a dredging excursion of the Berwickshire Naturalists' Club to Holy Island, off the northeast coast of Northumberland. He reported that "the whole of us were within an ace of being drowned." Nevertheless, the club members returned to the site a month later, for their anniversary meeting. This time, however, Johnson explained, "I took none of my family, in case of accidents!"[18] A measure of danger was appropriate for manly dredgers but rendered the outing unacceptable for women and children.

Because collecting was quite often organized through field clubs, dredging naturally took on the holiday atmosphere that characterized field meetings. Especially for Forbes and his dredging companions, excursions were characterized by a clubbishness evinced by inside jokes and songs. As he had for other occasions, Forbes composed a dredging song. Dredgers delighted in indulging in fishy puns, as when Forbes wrote to Alder that his research on starfish proceeded "swimmingly." Young American naturalists who worked under Spencer F. Baird's paternal direction at the Smithsonian Institution similarly formed a tight-knit group. William Stimpson, who relished the time he spent at sea, regaled Baird with his adventures, boasting on an 1861 excursion of plans for arming the schooner with a gun capable of firing [sea] shells so that, if the vessel encountered a privateer, they could "make them SEA STARS (N.B.—Joke)." From an expedition to the South Carolina coast the previous winter, Stimpson and his companion Theodore Gill wrote to Baird, each periodically wresting the paper from the other and adding his comments. Giddy with exhaustion from the hard work of dredging in numbingly cold water, they marveled at the zoological riches they had found, then turned to taunting each other that drinking, not dredging, made their hands so cold.[19]

It was not merely the element of outdoor adventure that spurred dredging activity, but especially the seagoing aspect. In the late eighteenth century, royalty and aristocracy had discovered the pleasure and salubrity of sea breezes and bathing. Seaside holidays, highly prized as social opportunities for the aspiring middle and upper middle classes, were embraced as delightful and healthy antidotes to busy urban life. In a similar way, sea voyages were considered salubrious, especially for young men suffering from the strains of studying too hard. Richard Henry Dana, for example, embarked on the voyage about which he wrote *Two Years before the Mast* in order to recover from the ill-effects of studying too hard.[20]

Naturalist-dredgers, too, felt that there was "no air beyond a sea breeze." Forbes, whose health suffered periodically after his Mediterranean voyage in HMS *Beacon,* found relief in ocean cruises. Describing to a colleague the benefit to geology if he went on a yachting cruise around Land's End to Southampton, Forbes also admitted, "I am very evidently not as well as I ought to be." His medical friends advised him to stay out of London, recommending the yachting excursion as an excellent way "to get pure air without much exercise." After the trip, he reported feeling much better.[21]

Serious scientific beachcombing and dredging excursions also took place during seaside holidays. For some, like Baird, the study of marine fauna was precipitated by the need of a family member to spend summers at the

Thomas Henry Huxley's sketch of a sail he made while honeymooning at Tenby, 1855. (From the Thomas Henry Huxley Papers in the Imperial College Archives, Notebook 125; courtesy of The Archives, Imperial College, London.)

shore for health reasons. Most, though, repaired to the coast for more purely social reasons. Thomas Henry Huxley honeymooned for several months at Tenby, a small beach town on the southern coast of Wales. While there, he spent his days collecting and studying marine animals. Sometimes he worked on specimens brought to him by others, but frequently he collected his own materials. Because he sometimes noted that the weather was too calm for dredging, he must have worked from a sailboat. A cartoon sketch on the cover of his journal depicts a man and woman in a sailboat, the man fending off an attack from a giant crustacean. Huxley's wife, Henrietta Anne Heathorn, had emigrated from Australia to marry him. After that long voyage, yachting excursions along the coast may have seemed rather tame, but because women were quite welcome on yachts, his bride probably accompanied him on some of his sea excursions.[22]

Huxley's honeymoon combined serious scientific work with a family social activity. In his book, *Depths of the Sea,* Charles Wyville Thomson commented on the satisfaction that scientists derived from dredging excursions precisely because they "combin[ed] the pursuit of knowledge with the recreation of their summer holidays."[23] Dredging trips were particularly compatible with family beach vacations. Not only were women welcome on pleasure craft, at least when the weather cooperated, but the beach rambles and bathing occupied children as well. While vacationing in Arran, a favorite resort of Glasgow residents, William Benjamin Carpenter remarked to Huxley, "You would be amused at the little bit of a place into which Mrs. Carpenter and I and our two boys are packed."[24] Many dredgers brought their older children along, even when they traveled to remote parts of Scotland, Ireland, or the outer islands. John Gwyn Jeffreys's second daughter accompanied him on a dredging holiday in Lerwick, in the Shetland Islands, where she occupied herself drawing, reading Italian, and collecting seaweeds.[25]

As middle-class access to the shore broadened, more opportunities for pursuing marine science appeared. Beachcombing and aquarium-keeping stimulated the development of marine zoology, but its organization through the British Association for the Advancement of Science transformed natural-history dredging into a national project.

WHEN THE TWO-YEAR-OLD British Association for the Advancement of Science met in Edinburgh in 1834, naturalists bemoaned the lack of attention paid to marine zoology. One speaker chided sternly, "I need scarcely say how small is the number of individuals who have added anything recently to our knowledge of the fish even of our own seas, not withstanding the opportunities for doing so which daily present themselves to naturalists resident on the coast." Forbes attended his first BAAS meeting in 1836. Three years later, he and Goodsir mobilized the Association to form a dredging committee. The first of many, this and its successors galvanized zoologists to undertake coordinated investigations of of the British coasts.[26]

Nineteenth-century natural history was not the pastime of pottering gentlemen endlessly classifying things to no apparent purpose, with a few truly scientific giants like Charles Darwin emerging miraculously from their ranks. Marine zoology provides a particularly good example of an undertaking grounded in traditional natural history yet simultaneously at the forefront of new professional science. Naturalists carried a vivid image of Newtonian astronomy into their taxonomic work, secure in their belief that, from the facts they observed of similarities between living things, a set of unifying principles would soon emerge. Ongoing description and classification were integral to the goal of discerning natural laws. Darwin's evolutionary theory provided a new research program for taxonomists after 1859. For Darwin, classification must be based on observations derived from comparative anatomy, embryology, geographical distribution, and paleontology. Instead of concentrating on an organism's form, naturalists investigated its life history.[27]

As the known number of species increased dramatically, naturalists began to specialize in one or a few faunal groups. Interest in marine forms grew out of this tendency to specialize, as did entomology, ornithology, and other specialties. For each faunal group—such as crustacea, polyzoa, mollusca, tunicates, foraminifera, and sponges—naturalists investigated the history of a species in terms of geographic origin, migration, and relationship to fossil species. They also faced the challenge of distinguishing between adult and larval stages of organisms, which, if they had been discovered, had sometimes been christened with two different species names. Naturalists charted the geographic range of species and tried to correlate characteristics such as temperature, latitude, or depth with their presence or absence.

The pursuit of these larger goals required the practice of traditional natural history: naming, describing, classifying, arranging collections, exchanging specimens, and producing local lists of species. Jeffreys, who addressed zoogeographic and geological problems in his published work, devoted most of his daily energy and attention to discovering and naming new species. His correspondence almost never mentioned theoretical issues. Instead, he discussed the validity of other naturalists' species definitions, the exchange of specimens, news of other naturalists' activities, and the minutiae of sorting and preserving collections. Letters written during dredging trips were preoccupied with the weather, but they also provided species lists and reported the day-to-day progress of the voyage.[28]

Zoology expanded rapidly in Britain during the first half of the century, but did so in the absence of formal scientific institutions and without university or government support. Individual zoologists amassed private collections of specimens and published species lists and descriptions in the burgeoning number of natural-history periodicals. They collaborated, visited each other's collections, recruited new members, and exchanged specimens, books, and ideas. Through extensive correspondence, they forged communities of specialists who provided the support functions for natural history that formal institutions did for other sciences.[29]

Marine zoology owed its vigor to networks forged through correspondence such as that between Joshua Alder and Alfred Merle Norman. Alder, a prosperous businessman for whom natural history became his life's work, expanded his interest in conchology to encompass all the phyla of northern European invertebrates, amassing an enormous and important collection. Active in local natural-history societies, he had an international scientific reputation and was a valuable resource to would-be marine invertebrate zoologists. Although Alder dredged occasionally when he was young, he stayed ashore when he got older, waiting for younger naturalists to send him specimens. In contrast, his younger colleague, the Reverend Alfred Merle Norman, participated eagerly in dredging excursions as often as he could spare the time from his parish activities. Like Alder, Norman had broad interests, publishing systematic work on polyzoa, crustacea, mollusca, tunicata, foraminifera, and sponges. He also built up a famous collection of 10,000 specimens. Called by a colleague "the greatest naturalist dredger of his day," Norman became the leader of local dredgers and a stalwart of the national community of marine zoologists.[30]

Within the British Association for the Advancement of Science, interest in dredging came not only from zoologists but also from geologists, who sought help learning about the present distribution of marine fossil fauna. The American zoologist Alexander Agassiz noted that scientists' ability to reach abyssal depths "shed [light] on many vexed problems concerning geographical distribution of animals and plants and their succession in time . . . and the problems concerning the formation of continents and ocean basins." From his perspective, in the latter third of the nineteenth century, he concluded, "We are now able to look back into the past history of the world with more confidence than heretofore."[31] When geologists began trying to make sense of mollusk species found in fossil strata, they encountered a dearth of knowledge. The meager information about mollusks that natural history could offer geology dwarfed available information on other marine forms, including echinoderms and crustaceans. During the 1830s, geologists recognized the importance of knowing the present distribution of fossil species, as when Charles Lyell enumerated for Section D of the BAAS the problems for geology of insufficient knowledge of marine zoology.[32]

Forbes encountered geologists' curiosity during the summer of 1839, when he dredged with the Scottish geologist James Smith in the Firth of Clyde and off the north coast of Ireland. During the 1830s Smith had observed raised beaches and exposed fossiliferous strata on Scotland's west coast. Either land and sea levels had changed quite recently, explaining why fossil species were still found living, or a longer geological time lapse explained why others were missing. Smith theorized that the missing fossils had lived in a colder climate, so he looked for these species in collections of northern shells. When he found them as living northern species, he argued that the British climate had been colder in the newer Pliocene Epoch and that the fauna had migrated northward, leaving the impression that species had gone extinct on the Scottish coast. Forbes's voyage with Smith not only acquainted him with the potential dividends that dredging offered geology, but also suggested to him the worth of differentiating between depth zones, leading to the formulation of his azoic theory.[33]

Such personal connections fueled the growth of dredging, and Forbes personally recruited many members of the original dredging committee formed at the 1839 British Association meeting. Most lived far from London, and many had access to vessels and acquaintance with local fishermen or other watermen. The Irish naturalists Robert Ball and William Thompson dredged with Forbes in the Killeries, a saltwater fjord on Ireland's Atlantic coast. Both were well known in BAAS dredging circles but not in the wider scientific community. Thompson, a Belfast merchant, was a good friend of Forbes and sailed and dredged with him on HMS *Beacon* in 1841. Ball, a government clerk in Dublin, invented what became the most popular version of the naturalists' dredge, which he exhibited at the 1849 BAAS meeting. Another Irishman from Belfast and a friend of Thompson's was Robert Patterson, who served on virtually every dredging committee through the 1850s. From the hotbed of marine zoology in Northumberland, the physician George Johnston served on the original committee. He published extensively on marine fauna, even writing an introductory conchology text. The yachtsman-geologist James Smith also participated. Few members did not fit the general profile, that of the provincial, amateur naturalist. One notable exception, John Edward Gray, a BAAS officer and Assistant Keeper of Natural History at the British Museum, seems to have chaired the first committee owing to Forbes's and Goodsir's youth and inexperience, but he dropped out the following year.[34]

Forbes, Goodsir, and other dredgers formalized what was already happening informally. Dredgers turned to the BAAS for both organizational and financial support. The actual fieldwork was organized locally, often through natural-history societies or field clubs. The committees identified desirable areas to investigate, introduced zoologists to potential collaborators, and provided subsidies. Just as manipulating dredges lay beyond the expertise of most naturalists, so too the expense of hiring vessels was out of reach for most. Dredging committees received small grants, usually about £25. Although the grants only partially defrayed the cost, they represented an important resource within the zoological community. The BAAS itself was founded largely in response to a sense among British men of science that their government, compared to other European governments, provided insufficient support for science. The dredging grants were the first BAAS funding of zoology. The support was mainly intended for actual collecting, although sometimes grants subsidized reports or illustrations. In total the British Association voted £1,440 to dredging committees over more than three decades.[35]

When Forbes reported his dredging results in 1850, he recognized England's "eastern coasts" as waters under the scrutiny of an active community of dredgers. Early in the nineteenth century, Newcastle-upon-Tyne and the surrounding area developed a strong regional focus on natural history, specifically marine zoology. The Natural History Society of Northumberland, Durham, and Newcastle-upon-Tyne served as a clearinghouse for regional activities from its founding in 1829. Among its creators were two central British Association dredging committee figures, Joshua Alder and Albany Hancock. In 1846, they helped found the Tyneside Naturalists' Field Club to organize field study and collecting trips. Through it they passed along their knowledge of, and interest in, mollusks and other marine fauna to younger naturalists. Tyneside's field club was inspired directly by the Berwickshire Field Club, inaugurated in 1831 with the novel feature of holding all-day meetings out of doors, in different parts of the district. The Berwickshire club, founded by George Johnston, descended from the popular field classes at Edinburgh in the 1820s.[36]

Northumerland dredgers enjoyed the advantage of proximity to the Dogger Bank, well known for its prolific fisheries but also a profitable ground for marine naturalists. Each summer from 1861 to 1864, BAAS grants helped defray the costs of dredging the Bank. The Tyneside Naturalists' Field Club arranged these excursions, in conjunction with the Natural History Society of Northumberland, Durham, and Newcastle-upon-Tyne. Members of the two organizations made up the large shortfall of cost after receiving the Association grant. They also carried out the dredging work, studied the specimens, and published species lists and papers in local scientific journals. In 1863, when the British Association met at Newcastle-upon-Tyne, many local dredgers took advantage of the opportunity to present their work. Alder described a new British polyzoan dredged off Shetland that year by Norman. Norman gave one paper on starfish and another on holothurians, a subgroup of echinoderms that includes sea cucumbers. George S. Brady presented a paper about marine cyclopoid entromostraca, small copepod crustaceans. His specimens came from both local dredging and a trip by Norman to the Shetland Islands. His brother, Henry B. Brady, described foraminifera, single-celled organisms with shells, that were new to the British fauna collected from locally dredged specimens. Henry later went on to write the volume of the *Challenger* report on that family. As a result of such active communities, marine zoology became a regional specialty across the northeastern part of England and the southeasternmost part of Scotland, as well as in Dublin and Belfast.

Pleasure was an essential ingredient for the dredging committees and local organizers because, for most of them, zoology was an avocation. Forbes acknowledged the large number of dredgers in Northumberland who did not get involved in central organizing efforts, including Richard Howse, Lieutenant Thomas, RN, Mr. W. King, Dennis Emberton, and Tuffen West. Another locally active dredger, George Hodge, supervised the operations off the Northumberland and Durham coasts. A businessman in Seaham Harbour, Hodge promoted his town's natural-history club and served as secretary to the regional natural-history society. He studied the marine zoology of local coasts, but not venture farther afield.[37] Even many of the leaders of British dredging were amateurs. Jeffreys, who chaired virtually every central and regional dredging committee for about a decade, quit his practice as a solicitor in 1866 to devote his attention to science. In Northumberland, Alder retired early from business to focus on conchology. Norman and Thomas Hincks were both clergymen, and George Brady was a physician. Albany Hancock started as a solicitor, but later he joined a manufacturing firm. Other active dredgers in the area were Henry Brady, a pharmaceutical chemist; George Johnston, a physician; Tuffen West, an artist; and Henry Tuke Mennell, a tea merchant. Elsewhere in Britain, groups of dredgers had similar profiles. Many northern Irish naturalists, including William Thomspon, were engaged in the linen trade. Also from Belfast, George Hyndman was an auctioneer and valuator. Edward Waller was a landowner. Doctors and clergymen had training in the natural sciences and jobs traditionally associated with amateur scientific pursuits. Merchants and other businessmen looked to natural history as rational entertainment, an appropriate channel for their energies.[38]

The dredging network also recruited career watermen to the cause. Jeffreys recommended W. Laughrin, an old coast guard man who became an associate of the Linnean Society, for the position of dredger and sifter on HMS *Porcupine*. Dredging committees sometimes hired working-class naturalists to help prepare reports. Charles William Peach was one such paid helper, who also became a well-known geologist and naturalist in his own right, serving on several dredging committees. Wealthy men of science such as Joshua Alder frequently hired him to collect for them. Like Laughrin, Peach was a coast guard man, stationed in Norfolk from 1824 until 1845, when he moved to Cornwall and began working as a customs officer until he retired in 1861. He moved in 1849 to Peterhead, a town north of Aberdeen on the east coast of Scotland, and then still further north to Wick. Fellow dredger Edward Waller said of Peach that he "had a general knowledge of most of the departments and is very strong in Lepralia and its more aspiring allies. Examines all stones and shells most anxiously." Waller also noted, "You know he is very chatty and attentive to the ladies and has his microscope to show off all the *wonders.*"[39] Peach established a scientific reputation for himself, publishing seventy-one scientific papers, receiving the Neill Medal from Edinburgh's Royal Society in 1875, and becoming famous for his discovery of Devonian fish and plants as well as Silurian lower invertebrates.

Participation in midcentury marine zoology cut across class lines, and practitioners also included a few of the first generation of professional scientists. The term "scientist" originated only in the 1830s, coined by William Whewell.[40] The differentiation of natural history into separate branches such as geology accompanied the professionalization of science, although accomplished amateurs could still earn the respect of their peers until late in the nineteenth century. Charles Darwin, to cite a notable example, did not earn his living by science. Only a handful of dredgers represented the increasing number of professional scientists populating the universities. Forbes cobbled together a variety of science-related jobs before winning the natural-history professorship at Edinburgh. Two professors, Dickie and Charles Wyville Thomson, joined Belfast-area dredging jaunts. The Dublin dredgers included three professors, E. Perceval Wright, J. R. Greene, and Melville, as well as two physicians, Kinahan and Carte.

Marine zoologists who earned their living as such grew convinced that experts must guide the field. Of all the scientific aims of nineteenth-century naturalists, geographical distribution became the central rallying cry for marine zoology. Characterized by none of the traditional disciplinary institutions such as journals, societies, and university chairs, the study of geographic distribution had no strong social or conceptual boundaries between scientific fields. The challenge of geographic distribution fascinated Forbes from the earliest days of his dredging activities in 1833. He and other naturalists of the time reacted against the habit of traditional naturalists of seeking only novelty or "monstrosities and organisms contrary to the law of nature." Forbes created blank forms to help naturalist-dredgers record consistent information. In doing so, Forbes drew on a tradition of natural-history instruction manuals whereby authors taught neophyte naturalists what kinds of perishable information about specimens must be collected along with the animal. Although he borrowed the format, Forbes had a greater ambition. He expected information from dredging papers to contribute to his project of mapping the presence and absence of specimens in particular areas as well as by depth.[41]

Spencer Baird also worried about unsystematic collection practices and warned his collectors against overlooking the most numerous species of an area, explaining that these might be rarities elsewhere. Baird's dissatisfaction with relying on untrained government surveyors and military officers led him to initiate a cogent program of scientific collecting in the 1850s. Baird communicated to the marine invertebrate zoologists William Stimpson and William Healey Dall his belief that control by a trained naturalist rendered the resulting collections capable of supporting taxonomic studies as well as investigations of speciation and variation.[42]

Concern with geographic distribution drove leaders of the British dredging community to expand operations. The Association met in different cities each year, all over the British Isles, which made it an effective venue for organizing such peripatetic work. A large, active group of naturalists began exploring the coasts of Ireland during the 1850s, initially working near Belfast and Dublin but eventually covering most of the shoreline. In the early 1860s, naturalists in Durham and Northumberland organized through the BAAS, although marine natural history had long been popular there. At the same time, a small group began dredging the Scottish coasts near Aberdeen, extending their work around from the east to the north coast of Scotland. Another group, from Liverpool, undertook dredging in the Mersey and Dee rivers. Until the 1860s, the centers of activity were far from London. Almost no work was done in the southern North Sea, the eastern English Channel, or the Thames River estuary.[43] These waters served as the cruising grounds for fashionable yachtsmen, but not naturalist-dredgers. By the 1860s, most of Britain's coasts had been explored by British Association dredging committees.

The meetings and field excursions put marine zoologists in regular contact with each other, encounters that fostered friendships and facilitated work. Leaders such as Alder and Norman served as clearinghouses for information on planning summer collecting trips. They helped neophyte naturalists find berths on dredging excursions, suggested scientific companions for dredging-yachtsmen, and advised each other on working with fishermen and local watermen in areas where they collected. Dredgers also sought advice from each other on the best localities for finding particular species, where to acquire dredges, and how to deploy them.[44] Face-to-face interactions were crucial for teaching dredging technique. Robert Ball liked to scatter pence over drawing room floors to demonstrate the correct position of the dredge by picking them up.[45] There was no substitute for going to sea, though, for teaching newcomers how to throw a dredge over board, how to judge when it hit bottom and when it was full, and how to haul it back in without dumping its contents in the process. Managing the labor of hired hands, necessary when dredging in depths of more than just a few fathoms, was likewise a skill best learned in person. Dredging fit well into the local scientific culture because outings offered occasions for exploration, adventure, and scientific neighborliness. Like BAAS meetings themselves, summer dredging excursions provided a combination of science and holiday-making that attracted midcentury naturalists and their families.

ALTHOUGH PIONEER dredgers worked first from rowboats or fishing vessels, they eagerly transferred their operations to the decks of yachts. Yachting provided a means for gathering marine fauna from the increasingly deep water in which collectors sought unusually beautiful shells and naturalists searched for undiscovered species. Yachts provided a more comfortable platform, both physically and socially, from which to practice marine natural history. The widespread popularity of yachting among the upper middle classes ensured that dredging gained many new adherents, varying widely in their devotion to science, but unwilling to endure fishing trips in order to catch glimpses of the unknown world beneath the waves. Dredging from yachts combined an enjoyable outing with the satisfaction of engaging in an uplifting and healthy activity that yielded rich hauls.

Yachting offered obvious advantages to naturalist-dredgers. Besides providing access to deeper water than rowboats, yachts allowed naturalists greater autonomy than fishing vessels to decide where, when, and how often to let down the dredge. Cruises allowed small groups of dredgers to learn or improve their technique or to experiment with different ways of rigging a dredge. They provided a focal point for holidays and a setting for naturalists' fish stories as well as scientific debates. During dredging holidays, zoologists planned future excursions and collaborated on research for publication. In short, yachting provided marine zoologists with much more than a physical platform for their collecting work. It allowed marine zoology to grow within the existing social structures that served natural history so well.

Before the nineteenth century, yachting was almost exclusively a royal activity. Yachts originated in seventeenth-century Holland, where they served as vessels of state, carrying royalty and ambassadors or participating in pageants and reviews. The same functions characterized English yachting after Charles II brought the idea of it from Holland when he ascended the British throne in 1660. Marine excursions became a favorite royal pastime, as did racing and betting. Dutch settlers brought yachting to New York and pleasure craft sailed in New York Harbor during the eighteenth century. Britain's first formal yachting groups appeared at that time: the Irish Cork Harbor Water Club in 1720 and the Thames's Cumberland Fleet in 1755. Limiting club activities to boats never let out to hire, these fleets engaged in squadron exercises under the orders of their commodore, reenacting famous naval battles or practicing complicated fleet maneuvers. War stalled yachting in the early years of the nineteenth century, but it revived during the post-Waterloo peace.[46]

The movement of people of high social position toward the seacoast introduced leisured gentlemen to a new summer pastime, one that had formerly been the work of pilots. Upper-class embrace of yachting followed closely in the shadow of royal fashion, just as migration to coastal resorts had a few decades earlier. When Victoria ascended the throne in 1837, yachting acquired an ardent patron. She ordered a succession of yachts and used them to visit the Continent, travel to her summer home on the Isle of Wight, make day trips and cruises, and watch the Royal Yacht Squadron races at Cowes. Basking in royal attention, Cowes transformed itself, during the regatta week, into the location of one of the greatest social occasions of the fashionable year.[47]

The first of the many nineteenth-century yacht clubs was founded in 1815 and received the appellation "royal" in 1820. Yachtsmen of the Royal Yacht Squadron (RYS) spent the club's early years cruising close to shore, attending annual festivals and processions, and holding a few small races. Activity up to that time focused on the utilitarian practice of naval squadron maneuvers and signaling, the displays of tactics under semi-naval discipline providing "spectacular aquatic exhibitions." By the mid-1820s, with French privateers subdued, racing around the Isle of Wight and saltwater cruising became popular, offering alternatives to traditional yachting practices. That decade saw the formation of several new clubs, mainly in northern England, Ireland, and Scotland. Clubs continued to proliferate through the 1840s; the first American club formed in New York City in 1844. These early clubs were exclusively the province of wealthy, prominent elites. In New York, yacht club members were considered even more socially exalted than the horse set.[48]

A new group of members joined the Royal Yacht Squadron at midcentury, those whose fortunes derived from great industrial and commercial enterprises. These wealthy social climbers had followed their aristocratic forerunners to the coastal resort towns before heading out to sea in their new yachts, intent on escaping the hordes of lower classes who populated the beach after the advent of rail travel in the 1840s. The activities of people who mattered socially were monitored closely by the press, whose reporters began attending annual yacht club meetings. A continual flow of comment on the doings of leisured yachtsmen and yachtswomen followed in the society papers. In an era when sports such as horse racing and boxing received extensive press coverage, so did yacht racing.[49] Yachting itself, and its society, became a spectator sport.

After midcentury, a band of "real sailors" began to disapprove of the dominant style of yachting, "a butterfly entertainment in which pretty faces and frocks were certainly as great attraction as the yachting." New millionaires vied with each other to build the fastest and most luxurious steam yachts, and it was this floating country-house living to which the new breed of yachtsmen objected. Sailors in small yachts made several heralded Atlantic crossings in the 1860s. In 1866, for instance, two Americans and a dog sailed a twenty-six-foot, two-and-a half-ton lifeboat from New York to England in thirty-four days.[50] Ocean cruising became popular with those who sought adventure, with people who loved, but respected, the sea. For them, yacht clubs facilitated practical arrangements for putting in at foreign ports rather than serving as the center of a social universe.

Not only were cruising yachts smaller and less expensive than luxury yachts, but their owners were more likely to be the kind of men whom pilots, tug skippers, and fishermen met on equal terms. In Britain, the provinces dominated serious yacht cruising, despite the numerical predominance of metropolitan yachtsmen. Most active naturalist-dredgers came from, or joined, the ranks of cruising yachtsmen. Yachtsmen-dredgers, like other cruising yachtsmen, were not necessarily from the highest levels of society. Although the monthly payroll in the 1870s and 1880s for the luxurious vessels of American millionaires such as John Pierpont Morgan or Jay Gould could run to $2,500, plus $1,500 for feeding the staff, the annual

Anne Brassey and her children welcoming shipwrecked sailors from the burning barque *Monkshaven* aboard their yacht *Sunbeam* in the South Atlantic, off Monte Video. (From Anne Brassey, *Around the World in the Yacht "Sunbeam"* [New York: Henry Holt, 1878], 115.)

cost of yachting was considerably less for average yachtsmen. In Britain, racing ran about £200 annually, but modest yachting could be enjoyed for £50 a year.[51]

Yachting developed into a major industry in Britain, providing natural-ist-dredgers with ready access to vessels and crew. In 1846, Britain boasted 530 yachts for a total of 25,000 tons, employing 4,700 men. These vessels carried 1,500 guns, a holdover from privateering days. By 1884, the number of yachts had increased to 2,300, of 110,000 tons, and employing 14,000 men. Paid yachting hands earned slightly less per week than men who gave gentlemen batting practice for cricket (30–50 shillings per week), although yacht hands were permitted to keep their clothing, worth about £7, at the end of the season. From the 1850s, most yacht owners ceded sailing duties to professional skippers. Yachting provided a major source of employment for residents of the coastal towns that became wealthy resort areas.[52]

Ocean cruising became a kind of do-it-yourself exploration. Anne Brassey's story about her family's cruise around the world in the yacht *Sunbeam* described keeping a daily log of position and air and surface temperature, rescuing sailors from a burning ship, capturing birds, weathering fierce storms, and other episodes typically found in narratives of exploring voyages. They even prepared a board with the yacht's name and the date to install on a Pacific island, to leave "a record of our visit."[53] Naturalist-dredgers searching for new species enjoyed the adventurous element of yachting as much as they appreciated the practical advantages of working from yachts. Some integrated scientific pursuits with travel for business and pleasure. Stripped of the excesses of elite yachting practice, cruising offered both strenuous exertion in healthy air and an opportunity to engage in the rational, uplifting business of scientific collecting. In short, yacht cruising provided an ideal context for dredgers who sought access to deep water.

MARINE ZOOLOGISTS were not the first men of science to use yachts as vehicles for furthering scientific knowledge. The Royal Yacht Squadron essentially functioned as a school for ship design. Naval officers swelled the ranks of the Squadron from 1816, when they were offered free membership unless they owned a yacht over ten tons. Men interested in shipbuilding as a science were drawn to yacht clubs, although their efforts sometimes went unappreciated, as did those of Thomas Assheton-Smith, a builder of steam yachts whose first few designs were not accepted by the Squadron. The BAAS served as a venue for reports on efforts to improve sailing vessels through the building and racing of yachts.[54]

Although most yachtsmen were content to leave ship design to others, many combined their travels with various sorts of naturalizing and geologizing. In 1846 the Royal Irish Yacht Club required owners to furnish the secretary with information about the coast and deep-sea fisheries wherever they cruised. Many yacht owners monitored advances in meteorology, collected measurements and observations, and followed the debates about oceanic circulation that began to rage in the middle decades of the nineteenth century. Much of this information had direct bearing on navigation in the waters they frequented. One of the Royal Yachts, *Osborne II*, became involved in the controversy about sea serpents that cropped up occasionally. In 1877, the captain and three officers witnessed a sea monster off the north coast of Sicily. The captain officially reported the sighting to the Admiralty and an officer published an account of the incident.[55] This sort of desultory scientific use of yachts was typical at a time when science was prized as a suitable occupation for leisured gentlemen who felt compelled to fill their time profitably. A group of serious naturalists, organized through informal networks and British Association dredging committees, began to employ yachts as tools for ocean science.

Forbes learned about the advantages of yachts for dredging from James Smith on the 1839 expedition where he encountered the connection between geology and dredging. Smith enjoyed substantial leisure as a sleeping partner of his father's West Indian trading firm. When he began cruising the Scottish west coast and the Irish Sea in 1806, he developed an interest in geology and archaeology, and his social circle grew to include the scientific Hooker family and the Antarctic explorer James Clark Ross. Through dredging Smith sought to prove his theory that species of marine fauna recorded in the fossil record but absent from British waters had lived in a colder climate than that prevailing at the present. His scientific work won him election to fellowship in the Geological and Wernerian societies.[56]

Yachting permitted naturalists to organize and conduct serious scientific research individually and in small groups, utilizing the informal personal networks through which they were accustomed to exchanging specimens and information. As early as 1841, dredgers with access to yachts, either their own or a friend's, began working from them. That year George C. Hyndman, of the Belfast Natural History Society, reported on deep dredgings made off Sana Island while cruising with his friend Edmund Getty, in the yacht *Gannet* on the nineteenth of July. They employed the dredge three times that day in forty fathoms and concluded that the area corresponded to Forbes's "coralline" zone. Hyndman reported a rich haul, listing seventy species obtained. In 1857, he thanked several gentlemen amateurs for lending yachts and providing assistance for the work of the Belfast dredging committees over the years. He particularly noted the addition, in 1856, of Edward Waller, who took a party of naturalists including Hyndman and Professor George Dickie out in his yacht to a submarine bank lying at twenty-five to thirty fathoms. Dredgers discovered this bank in 1852 but, as Hyndman admitted, it was then already well known to fishermen as Turbot Bank. Waller's dredging excursion yielded sand so rich in unusual shells that he sought the help of the expert conchologist Joshua Alder to compile a species list.[57]

One of the most energetic dredgers of the first half of the century was Robert MacAndrew, christened by Forbes "that indefatigable friend of submarine research." MacAndrew was a Liverpool fruit merchant whose scientific interest in mollusca developed out of shell collecting. His father, a London businessman who came from Scotland, died in 1821, when MacAndrew was nineteen, so he inherited his share of the business and the responsibility at a young age. Eight years later, following the death of his brother, he moved to Liverpool, where his brother had lived. Like many dredgers, MacAndrew turned to dredging to expand his collecting hobby. He began with an open boat, then used a small sailing boat. Finally he outfitted two yachts for cruising and dredging. MacAndrew embraced yachting in order to dredge rather than turning to dredging to occupy himself while at sea, as Smith had. He traveled frequently, especially to Spain for business, and his collection grew. In 1834 he joined the Liverpool Literary and Philosophical Society to meet other scientists.[58]

As with so many others, Forbes introduced MacAndrew to the dredging community and got him involved in BAAS committees. MacAndrew met Forbes either at the 1837 BAAS meeting in Liverpool or in 1840, when Forbes returned there to deliver a series of eight lectures. MacAndrew began his long-standing work with fellow British Association dredgers in 1842, when he received a modest grant of £5. He won larger grants each of the next two years (£50 and £25), and then became a member of the dredging committee in 1844. After dredging on the western coast of Scotland that summer, he cowrote the report with Forbes. In 1850 Forbes wrote a major overview of the progress in marine zoology based on a decade of dredging. He decided to limit the species lists to results from dredging papers that had, with one exception, been filled out "on the spot" by himself and MacAndrew. He cited his concern that not all shores had been explored "in an equally systematic manner," noting that, "well-filled tabulated forms [were] still wanting" from other naturalists.[59] His confidence in MacAndrew's work stemmed from the fact that he frequently accompanied MacAndrew on cruises and had personally taught him how to fill out the dredging papers.

MacAndrew outfitted his own yacht, the *Naiad,* and set sail in 1849 on a five-month cruise to Spain, Portugal, Malta, and southern Italy. The trip had the dual purpose of improving the health of his sixteen-year-old son, William Edward, and collecting shells. MacAndrew came from a large, close-knit family; he married a cousin and shared a business with relatives. He often took along his wife and some children (they had eleven, of whom eight lived to adulthood) when he traveled. He was joined on this cruise by his wife, three sons, two daughters, and a girl cousin. On a voyage two years later around Scotland, he brought his wife, two sons, and two daughters. A third jaunt in 1855 found MacAndrew with one daughter and one son, sailing and dredging around Shetland and up the Norwegian coast.

These trips yielded specimens that constituted one of the finest shell collections in Britain. They also provided MacAndrew with an opportunity to explore deeper waters than most naturalist-dredgers had. While off Norway, MacAndrew dredged four hours a day, except Sundays, collecting regularly from 100 to 200 fathoms. One attraction of northern areas such as the north coast of Scotland, Shetland, and Norway was the existence of deep water in close proximity to land. Soon after he returned, he predicted that he would find six to eight new species in his hauls.[60]

Between these longer cruises, MacAndrew dredged with scientific friends closer to home. Just after the Edinburgh British Association meeting in 1850, Forbes and his friend John Goodsir, his collecting partner from his student days in Edinburgh, dredged with MacAndrew for three weeks in the deep-sea banks off the Hebrides Islands. While he cruised on the *Naiad,* MacAndrew's name disappeared from the dredging committees lists, but in 1858 he reemerged to head the new General Dredging Committee, established to orchestrate the activities of the proliferating local committees. This position acknowledged the leadership role he played. Not only did MacAndrew become a well-known collector, but he also earned a good reputation for his scientific work. He published papers in the *Proceedings of the Literary and Philosophical Society of Liverpool,* the *Annals of Natural History,* and the *Athenaeum.* In 1847, he was elected a Fellow of the Linnean Society and, in 1853, of the Royal Society. When Forbes's book, *The Natural History of the European Seas,* was published posthumously in 1856, Robert Godwin-Austen acknowledged MacAndrew's work again: "We should hardly have ventured, even now, to speak thus confidently of the relations of the Mediterranean fauna, but for the recent researches of our own countryman, Mr. MacAndrew." MacAndrew retained the General Dredging Committee chair until 1861, when it passed to another yachting-naturalist, Jeffreys.[61]

Jeffreys dominated British Association dredging activities in the 1860s, the same decade he wrote the definitive five-volume work *British Conchology.* He was also the most active yachtsman-dredger. His scientific interest began early and by the age of nineteen he had published his first paper. The following year he was elected to the Linnean Society, and fellowship in the Royal Society followed in 1840. A native of Wales, Jeffreys spent his holidays dredging from a rowboat in Swansea Bay, more interested in science than in his training to be a solicitor. In 1856, he moved to London to practice law, but his main motivation was to be closer to the scientific community there. Soon after, he began to devote most of his time to conchology. In 1866 he retired and bought a home called Ware Priory in Hertfordshire, partly to house his growing collections.[62]

Jeffreys began working farther from Swansea Bay in the 1840s. In 1841 and 1848, he dredged off Shetland, basing his operations in the town of Lerwick. In 1846, he dredged around Skye. There is no record that he ever dredged with Forbes and his name first appeared on dredging committee lists in 1858, four years after Forbes's untimely death. He did dredge with committee chair MacAndrew, although the two conducted an extended argument in print over MacAndrew's support for some of Forbes's ideas. Although Jeffreys came to agree with Forbes's description of zones with characteristic faunal assemblages, he believed that collecting from deeper water would prove that no species was limited to a particular area. Between 1861, when Jeffreys succeeded MacAndrew as the chair of the General Dredging Committee, and 1867, Jeffreys headed every single British Association dredging committee, including local ones formed to dredge Dogger Bank, the north and east coasts of Scotland, the Mersey and Dee rivers, and the Bay of Dublin.[63]

Although Jeffreys participated in debates about zoogeographic and geological problems, his primary motive for dredging appears to have been the search for new species.[64] The key for all these lines of inquiry, especially the latter, lay in acquiring specimens from ever greater depths. Starting in 1861, Jeffreys undertook an ambitious program of dredging around Shetland, mostly aboard a yacht he had borrowed, and eventually purchased, from his brother-in-law.

Shetland appealed to naturalists as a collection site because of its geography. Lying closer than the rest of the British Isles to deep water, where conditions resembled those of preglacial times, Shetland seemed a likely place to find "living fossils," creatures previously known only from the fossil record. The location was also ideal for exploring the connection between northern, or Arctic, fauna and British forms. More important, Shetland was not zoologically well explored. The few forays made by naturalists before the 1860s hinted at the riches waiting to be discovered. Forbes and Goodsir dredged there in 1839 and again, with MacAndrew, in 1845, finding over a dozen animals new to British lists. When Jeffreys collected there in 1841, he found twenty-one species new to the area and, in 1848, he discovered several living fossils. In 1857 and 1858, Jeffreys hired George Barlee, about whom little is known except that he was an active dredger who knew both Jeffreys and Forbes, to dredge for him in Shetland. The remote northern islands were so unfamiliar to his British Association audience that Jeffreys found it necessary to describe their geography, remarking that, "when it is recollected that Byron confounded the Orkneys with the Hebrides, . . . I trust I may be excused for saying a few words as to the position of these sea-girt isles."[65]

Jeffreys characterized his 1861 trip to Shetland as a "yachting excursion" with two scientific friends, but it and the rest of his northern trips were very much working holidays. In 1863 he received a BAAS grant that covered a quarter of the £300 that the trip cost. Each year that Jeffreys worked off Shetland or the Orkneys, the BAAS granted £50 to £75. This was more than other dredging committees received, because, in the words of one of Jeffreys's colleagues, work in open Atlantic waters entailed "not inconsiderable outlay." The costs were borne principally by Jeffreys and a Mr. John Leckenby who, Jeffreys noted gratefully, contributed liberally even when he could not accompany the expedition. Other participants gave "much smaller sums," but "not less willingly in proportion to their means."[66]

Jeffreys's yacht, the *Ospry*, was critical to his success. In 1862, and briefly in 1864 after *Osprey* was de-masted in a gale, Jeffreys chartered the *Gem*. In 1863, he engaged the steamer *Xantho* from Scarborough. Most summers, Edward Waller and Alfred Merle Norman accompanied him. Sometimes George J. Allman came along, or coast guard and customs officer Charles Peach. Scientific expertise was helpful, but the labor and technical skill of Jeffreys's hired dredgers was indispensable. In 1863 he employed Captain Phillips and five crew members to handle the boat and work the dredge. The following year, he sought greater expertise and hired Archibald McNab of Inverary to oversee the dredging operations. McNab had formerly worked for Jeffreys and Barlee, dredging on the west coast of Scotland. In a letter to Norman, Jeffreys praised his crew, "McNab is a capital fellow and quite up to his work. Phillips and his five men behave excellently and work with a will."[67]

Jeffreys experimented with techniques until he devised a system of deploying two dredges at a time, to maximize collections, which had to be squeezed in between the frequent storms and rough seas he encountered off the northern islands. This practice also guarded against the loss or malfunction of one dredge and yielded the scientific dividend of ensuring a representative haul. Although most BAAS dredgers adopted Ball's dredge, Jeffreys preferred bigger and heavier ones that more closely resembled their industrial precursors. In 1844 he confided to Joshua Alder, "I do not think much of Mr. Ball's toy dredges, although I took one with me (in addition to my own) last year for experiments in deep-sea dredging."[68]

Heavier dredges sank faster and were more likely to reach bottom in deep water. Large dredges, however, collected so much mud that the strain of hauling them in often broke the rope. Experience taught Jeffreys that small but heavy dredges worked best to reach great depths. Using his methods, his crew regularly reached depths at the extreme of those explored by naturalists working from yachts, in the 100- to 200-fathom range.[69]

Heavier dredges, plus longer lines to reach greater depths, taxed his crew, who often labored in extremely stormy conditions. The work was hardly more pleasant because of the mittens he provided to protect their hands while hauling in the dredges. The sailors deployed the dredges in the evening, leaving them out all night and hauling them aboard in the morning. This practice, along with the pressure to work while the weather allowed, frequently resulted in fourteen-hour days. Often the weather did not cooperate, as when Jeffreys reported to Norman that, in strong gales, "the men could not keep their feet while hauling up."[70]

McNab, the dredger, took an interest in the work and became skilled enough that Jeffreys could send the yacht out without having himself or one of the other naturalist-dredgers on it. Another of Jeffreys's hired dredgers had a long and varied career. W. Laughrin was "an old Coast Guard man" from Polperro with whom Jeffreys had worked before Jeffreys recommended him for a job on HMS *Porcupine* as dredger and sifter for its 1869 and 1870 cruises. Laughrin became an associate of the Linnean Society and a well-known member of the dredging community.[71] Most crew members did not, of course, become so involved in dredging as McNab or Laughrin, but naturalists depended as much on the labor pool generated by the yachting industry and other maritime occupations as on the yachts themselves.

Many aspects of the life and work of yachting naturalists such as John Gwyn Jeffreys, James Smith, and Robert MacAndrew were similar. All three loved to travel and embraced yachting to cruise in both British and foreign seas. All were wealthy enough to spend weeks, even months, at a time, yachting, traveling, and collecting. They outfitted their vessels with not only the usual appurtenances of gentlemanly life, but also the equipment needed to dredge in deep waters. To varying extents, they sought scientific companions to share the expense, labor, and faunal dividends of dredging. All three were "amateurs," although Smith and Jeffreys became respected men of science. MacAndrew was perhaps better known as a collector and adventurer who had an active interest in science. All three manifested the nineteenth-century tendency to turn to science as a morally appropriate pastime. The moral appeal of marine science combined with the sociability of yachting also invited women to participate.

JUST AS royal example promoted the sport initially, Queen Victoria's enthusiastic embrace of yachting set a precedent for women to go to sea. Especially in the 1860s and 1870s, ladies were invited into clubhouses at festivals and included on the day sails and rounds of lunches and parties that characterized fashionable yachting circles. They joined the ranks of spectators at races, but participation in cruising and day sailing was open to women. Queen Victoria, it was said, knew enough of seamen's language to communicate her wishes to her crew. The Royal Yachts had special ladies' apartments attended by serving women who were carried on the ships' ledgers as sailors with naval ratings. Lady Brassey lived for a year with her family, including her four small children, on their yacht *Sunbeam* as they sailed around the world. Although women in fashionable yachting society did not usually participate in the physical handling of yachts, other women did. Among younger couples, especially those who cruised geographically farther from the social centers of yachting, it was not uncommon for women to act as crew members .[72]

In dredging circles, women were quite welcome on collecting excursions. They frequently accompanied naturalists' dredging parties not just as companions but as participants. The 1871 Edinburgh meeting of the British Association featured a "great Dredging Expedition" of sixty naturalists off Bass Rock. The report named a handful of well-known scientists in the group and noted that there were "also a number of ladies in the party." In 1873, twenty members of the inland Birmingham Natural History and Microscopical Society, including several women, spent a week in Teignmouth dredging from the yacht *Ruby.* Afterward, participants proposed to make such an excursion annual, "especially as ladies were now for the first time admissible as members."[73] Because of the social aspect of yachting, sea voyages became not only possible, but quite acceptable, for well-bred women. This access opened up the possibility for women interested in marine natural history to participate in the oceangoing version of it.

And women did. Progress in marine zoology was solidly rooted, as Forbes noted, in the "diligent and searching investigations" of the many amateur naturalists. Some female naturalists worked with a father or husband; many had collegial relationships and correspondence with other naturalists. Margaret Gatty, for instance, collected and identified specimens for William Harvey, and also helped him with his correspondence. Dr. George Johnston urged women to join the full-day, outdoor meetings of the Berwickshire Naturalist's Club, always saying, "we must not forget the women." In his own family, his wife illustrated most of his published work, while his daughter translated manuscripts for the Ray Society.[74] Jeffreys also encouraged women to join the ranks of naturalists, enjoining them, "Don't ignore knowledge, nor be ashamed of using the intellect and faculties which God has entrusted to you . . . Don't say, 'Oh! I am not scientific,' . . . from hugging yourself with the consciousness of possessing

Women were welcome on yachts and participated in the sailing; here the woman at the tiller is learning to steer. (From *Frank Leslie's Ilustrated Newspaper*, July 15, 1871.)

some recondite virtue." In an address to the newly formed Hertfordshire Natural History Society and Field Club in 1879, he stressed ladies' participation in provincial societies and field clubs. He pointed out that half of the 500 members of the Liverpool Naturalists' Field Club were women and noted that many ladies took part in dredging expeditions to distant places organized by the Birmingham Society each year. In addition, Jeffreys took the concrete step of making his scientific work accessible to "lady readers" who had not studied classical languages by including accent marks over genus and species names to instruct readers in correct pronunciation.[75]

Not all branches of natural history were equally open to women, but marine zoology became one of the most accessible. Shooting and skinning were not considered appropriate activities for ladies, and thus women tended not to work in ornithology. Botany, with its complete lack of blood and guts was popular among women naturalists, although entomology gained ground in the early decades of the nineteenth century in spite of some qualms about the necessity of killing the animals. Marine zoology, which became very popular at midcentury, appealed to women for some of the same reasons it appealed to men. Collecting provided an opportunity for healthy outdoor exercise. It added a purposeful dimension to beach holidays. Shells had long been desirable cabinet acquisitions, but midcentury popular natural-history books enjoined men, women, and children to go to the shore and observe living specimens. The aquarium craze, which swept through Britain in the 1850s and hit the United States the following decade, enabled whole families to transport living marine fauna and flora into their homes.[76]

Several characteristics of marine zoology made it particularly appealing to women, especially in comparison to other branches of natural history. The moral ease with which naturalists could dispatch their captures certainly encouraged women to turn to marine fauna. Women's social access to the sea through bathing, holiday-making, and yachting may explain the large numbers of women who practiced marine natural history on an amateur level. By contrast, women were less likely to take up field sciences such as geology partly because they did not usually take part in mountaineering or other strenuous outdoor activities. The vogue of seaside holidays and the widespread acceptance of women in yacht-cruising circles supported the entrance of women into serious marine zoology.

Women did succeed in gaining recognition for research on marine fauna and flora. In Britain, one woman who became a marine zoologist at the Plymouth Laboratory of the Marine Biological Association, Marie V. Lebour, came from a neighborhood that produced numerous active naturalist-dredgers. Born in Northumberland in 1876, the daughter of a geology professor at Newcastle-upon-Tyne, Lebour earned all her degrees, including a D.Sc., from Durham University. She held positions at the Universities of Durham and Leeds until 1915, when she was lent to Plymouth for the duration of the war. She remained there for the rest of her career.[77]

In the United States, women also became established in marine zoology. For two years in the mid 1870s, Louis and Alexander Agassiz ran a summer school on Penikese Island, off the coast of Massachusetts. They admitted fifteen women to their first class of about fifty students, and this number increased to twenty the second year. The school offered these women the rare opportunity for advanced training in biological research. In spite of its short duration, Agassiz's school inspired the creation of other summer seaside teaching laboratories, most notably the Marine Biological Laboratory at Woods Hole (MBL) in Massachusetts in 1888. Three of the eight original trustees of the MBL were women committed to promoting educa-

Victorian men and women posing on a large rocky outcropping at the British coast, 1867. Although photography at the time did not allow for action shots, beach holidays often provided a setting for dabblers or serious naturalists to collect and observe seaweed and marine organisms. (Hulton Archives / Getty Images.)

tional opportunities for women, and summer classes at Woods Hole attracted many women, both teachers and research scientists.[78]

But women's access to ocean science was virtually eliminated when yachts were supplanted by different vessels for studying the sea. In 1868, British men of science embarked on HMS *Lightning,* whose voyage initiated a decade-long partnership between the Royal Society and the Admiralty devoted to studying the oceans. Once marine science began to take place from the decks of men-of-war, and soon thereafter cable ships and fisheries research vessels, women were excluded because these work settings were not open to them.

The preference ocean scientists developed for working vessels over pleasure craft was consistent with strategies used in other branches of science at the time to exclude amateurs and women. Women's exclusion from oceanography was not an unavoidable consequence of its origin on naval vessels. Natural science of the sea began not on such ships, but on rowboats, yachts, whalers, fishing vessels, passenger packets, and surveying vessels. Women's exclusion was part of the professionalization of the discipline, a by-product of efforts by ocean scientists to define themselves as such. They did so by transplanting their collecting activities into work settings to which professional scientists could control access. Ocean science conducted on yachts by vacationing naturalists gave way to oceanography practiced by professional scientists aboard working vessels.[79]

chapter five

Dredging the Moon

"That great highway extending from pole to pole, which is for ever closed to human gaze, but may, nevertheless, be penetrated by human intelligence."

—George C. Wallich, *The North Atlantic Sea-Bed,* 1862

The great depths of the ocean are entirely unknown to us . . . What passes in those remote depths—what beings live, or can live, twelve or fifteen miles beneath the surface of the waters—what is the organisation of these animals, we can scarcely conjecture.

—Pierre Aronnax, in Jules Verne's 20,000 *Leagues under the Sea,* 1869

MARINE ZOOLOGY began as the pastime of provincial naturalist-dredgers and a few devoted mollusk specialists. By the 1870s, though, John Gwyn Jeffreys could assert that the ocean was "a new world full of interest not only to the naturalist but to every man of science."[1] By then, professional scientists associated with important metropolitan scientific institutions—physical scientists in addition to zoologists—had taken notice of the deep ocean's potential for harboring important truths about nature.

During the 1860s, naturalists reached the limits of depths they could dredge from their yachts. George Barlee, a friend of Edward Forbes whom Jeffreys hired as a dredger, complained to Johsua Alder that it was "as well to dredge the moon as 50 or 70 fathoms with oars."[2] Yachts allowed Jeffreys to dredge regularly in 80 to 100 fathoms and occasionally deeper. Naturalists determined to reach greater depths began to hire steam yachts. These were even more effective for deep-water collecting because they allowed fast, reliable, and comfortable travel to offshore deep water. Such charters were expensive, however, too expensive for all but the most wealthy and devoted dredgers. In addition, steamers provided mechanical aid only for transportation of the naturalists and sailors; dredges were still deployed, and hauled in by, human labor.

Although most naturalists remained content to explore parts of the sea they could reach relatively easily, a few of them imagined the possibilities of collecting from the ocean's deepest waters. Naturalists had occasionally reached depths in the 200- to 300-fathom range when working from the decks of naval vessels, as when Forbes sailed on HMS *Beacon* to the Mediterranean in 1841–42. Ambitious zoologists, chafing at the limits of dredging from yachts, began to look for naval assistance and government patronage.

Success in attracting such help was fueled by the debate over whether life could exist at the ocean's greatest depths. Great fanfare surrounded the discovery in 1860 of marine animals encrusted on a Mediterranean submarine telegraph cable recovered from 1,700 fathoms. But zoologists' realistic expectation that the ocean's depths could be breached depended on the new metropolitan headquarters of ocean science. Although provincial naturalist-dredgers continued to investigate the coasts, London-based scientists grew interested in the secrets of the sea. Naturalists also found a sympathetic audience among those physical scientists who followed the great meteorological debates about the cause and nature of oceanic circulation. An active program of deep-sea sounding by the British Hydrographic Office prompted naval interest in ocean science. Hydrographers provided the expertise and equipment for pioneering deep-sea dredging.

Dredgers worked through existing social networks to take advantage of growing scientific, political, and public interest in the depths. Friendships, family connections, and social acquaintances provided avenues through which these spokesmen for ocean science organized cruises that satisfied marine zoology's emerging scientific goal of exploring vast depths. The cruises of HMS *Lightning* and HMS *Porcupine* and the more famous voyage of HMS *Challenger* in the late 1860s and early 1870s were the result of behind-the-scenes arrangements made long before official correspondence formalized cordial calls and friendly conversations. After just a few short years of lobbying, dredgers succeeded in establishing the depths as scientifically significant for both biological and physical science.

THE QUESTION of whether or not life could exist at great depths inspired the strongest interest in deep-sea dredging. Like near-shore dredgers, deep-ocean investigators wanted to find new species, study their geographic distribution, and search for "living fossils." Well into the nineteenth century, different populations held divergent opinions about the existence of life in the deepest parts of the ocean. For decades naturalists accepted the judgment of Edward Forbes—that animals could not survive in the physical conditions of depths below 300 fathoms. Even in the 1860s, some naturalists chose to interpret the changing evidence of life at great depths as simply stretching the limits of Forbes's azoic zone farther down into the sea.[3]

Long before the 1860s, hydrographers, deep-sea fishermen, and whalers had found evidence of living organisms at great depths, reports of which appeared in public channels such as mariner's publications, expedition reports and narratives, newspapers, and the periodical press. Coexistence of conflicting evidence indicates that naturalists and mariners lived in different intellectual and social worlds. That some mariners were gentlemen officers makes the point even stronger; gentlemen of science had little inclination to consider the observations and specimens collected by career seamen.[4]

The question about life in the depths also related intimately to the disagreement over whether the sounding machine or the dredge was the best tool for deep-sea investigation. Hydrographers and naturalists discussed this issue more often than the scientific problems that the technology was intended to solve—the effects of extreme pressures and temperatures on deep-sea life. Naturalists favored the dredge, while hydrographers insisted that a modified sounding machine was a more appropriate tool. Initially, naturalists examined whatever material hydrographers could recover from the ocean floor. Because sounding machines were the only equipment to reach below a few hundred fathoms before the late 1860s, they served as zoological sampling devices for microscopists. As the final arbiters of the best equipment for investigating oceanic life, however, zoologists chose the dredge.[5]

At midcentury, virtually all naturalists concurred with Forbes's theory that life reached a zero at some depth in the ocean. This belief proved marvelously persistent in the face of mounting evidence to the contrary, evidence most naturalists simply did not acknowledge. In 1819, Sir John Ross reported finding worms alive in a 1,000-fathom sounding, and starfish attached to a sounding line at 800 fathoms. Unfortunately, his expedition carried no naturalist, and the specimens did not arrive in England in a state fit for identification and description. During his 1829–30 expedition, Ross failed to return with zoological specimens from his dredgings because he abandoned the collection along with his ship, the *Victory*. Ross's nephew, Sir James Clark Ross, seemed determined to succeed where his uncle had failed. He brought dredges and a naturalist, the young Joseph Hooker, along on his 1839–1843 expedition in the *Erebus* and the *Terror*. Their deepest dredging reached to 400 fathoms and several other times they dredged to 200 to 300 fathoms. But after Ross returned he let the specimens decay unexamined. In the mid-1840s, others successfully dredged in the 300-fathom range, including Harry Goodsir, the naturalist of the doomed Franklin expedition, as well as Captain Thomas Spratt, who worked for the Hydrographic Office. Goodsir perished in the Arctic with the rest of the expedition, so naturalists had only his letter describing the dredge haul, not even a formal species list.[6]

Many mariners, especially hydrographers, shared the opinion of James Ross that life would be found at all depths. Ross acknowledged frankly that his opinion differed from accepted scientific wisdom: "Contrary to the general belief of naturalists, I have no doubt that from however great a depth we may be enabled to bring up the mud and stones of the bed of the ocean, we shall find them teeming with animal life; the extreme pressure at the greatest depth does not appear to affect these creatures."[7] Observations by whalers such as William Scoresby reinforced such convictions.

When naturalists faced choosing between, on the one hand, the opinion of a respected naturalist such as Forbes who had done extensive field research and, on the other hand, explorers who did not report using the practices and publication outlets preferred by naturalists, they believed Forbes. His dredgings in the Mediterranean from the *Beacon* clearly demonstrated a diminution of life with depth, with no specimens collected below 300 fathoms. Forbes presented his results in scientific papers and publications, including a book that he was writing when he died in 1854.[8] Explorers, by contrast, announced their zoological finds casually in voyage narratives or sometimes as appendices to official reports to the Admiralty. Their claims were neither aimed at the community of naturalists nor presented in the language and format that naturalists recognized as scientific. The fantastic stories of explorers were given no credence in taxonomies and natural histories.

Naturalists resisted such tales about life at great depths in large part because no specimens surfaced to support them. Specimens acted as a kind of currency among the community of naturalists. Before institutions such as the British Museum took over responsible curatorship of specimens, they circulated among naturalists along with correspondence. Until after midcentury, these networks performed many of the functions traditionally attributed to scientific institutions, including recruitment and training of new members. They also facilitated the accumulation of large collections that were essentially held in trust for the zoological community.[9]

Naturalists' insistence on the availability of specimens to support claims of new species makes it evident why reports by hydrographers of life at great depths passed unremarked. A parallel case exists for nineteenth-century scientific sea monster sightings. Such reports increased dramatically as geological exploration began to recover fossil evidence of fantastic ancient monsters. Discovery of plesiosaur fossils made the possibility of present-day sea serpents seem probable. Sea serpent investigation was treated seriously, if cautiously, by men of science. In 1817, members of the Linnean Society in the United States investigated sightings near Gloucester, Massachusetts. Lacking specimens, the naturalists relied on testimony from witnesses who were naval officers and crew, whalers, and fishermen. In the first flush of excitement, they found enough evidence to name not only a new species, but also a new genus. This accomplishment appeared a testament to the bounty of American nature, the heroism of American fishermen and whalers, who were shortly expected to capture the beast, and the genius of American science. Expeditions sent the following year to hunt the monster not only failed to capture it but also called into question the validity of the sightings.[10]

This episode did not dissuade men of science, including Louis Agassiz and Charles Lyell, from investigating sightings. Lyell believed it unlikely that modern serpents would be reptiles, but thought a lack of evidence was an insufficient reason to rule out their existence. Even he marginalized his own work on sea monsters, though, revealing his interest only in a private notebook. Richard Owen, head of the natural-history department of the British Museum, also collected information on sightings, although he discredited such investigation by calling attention to the lack of specimens or remains of any such creatures. By midcentury, American naturalists were firmly convinced that most sightings had other explanations. During the North Pacific Exploring Expedition, a shore party saw "what appeared to be a succession of milk-white coils, simultaneously rolling, and a strong opinion prevailed for the moment that the Great Sea Serpent was approaching." The expedition's naturalist, William Stimpson, diffused the excitement by declaring the "serpent" to be merely several beluga whales swimming together.[11]

Stimpson's mentor, Spencer Fullerton Baird, collected notices of sightings and captures of purported sea serpents, and corresponded with other naturalists, including Louis Agassiz, about them. Baird maintained a skeptical attitude, confiding in one case that "the account seems to be veritable and it is a great puzzle," only to be relieved a few days later by an update that the "monster" had proved to be a strange fish. As late as 1883, though, Baird asked keepers of Life-Saving Stations, who were responsible for collecting stranded marine mammals for the Fish Commission, also to look out for "the great basking or bone shark, and any unknown or unidentified marine monsters, such as might possibly suggest the idea of the far-famed 'sea-serpent.'"[12] Baird and other naturalists were more interested in the nature of unknown and unexpected oceanic fauna than in whether or not a great sea monster existed. But naturalists needed specimens, not sightings.

Like sea serpent investigators, naturalists who addressed the question of the existence of life at great depths labored under the burden of negative evidence. All of the stories of animals pulled from the depths before 1860 had in common that the specimens, when any were collected and preserved, did not surface inside the community of marine zoologists. The fact that Sir James Clark Ross preserved deep-sea specimens suggests that he recognized their importance to naturalists. Unfortunately, as the *Challenger* narrative authors later lamented, these collections were not described and were therefore "lost to sight."[13]

But the year 1860 saw two discoveries in quick succession that did attract the attention of naturalists. In both cases, specimens made their way into the hands of established zoologists. One discovery was made by George C. Wallich, one of the few naturalists who took seriously information about the ocean derived from seamen. Wallich embarked on the *Bulldog* fully expecting his findings to vindicate his confidence that animals could exist at great depths. A week into the trip, he had already received some confirmation of his views. From a 680-fathom sounding, he reported living foraminifera, a serpula (a marine worm), an annelid, and a fish attached to a piece of granite rock. At depths slightly less than 680 fathoms, he found annelids and one crustacean. He confided to his journal, "If I can only get the proper opportunities, I have not a vestige of doubt that I shall be able to prove the presence of life at all depths."[14]

Wallich relied more on sounding machines than on the traditional naturalists' tool, the dredge. *Bulldog* officers used several different sounding devices, including a clamlike grab sounder that collected much more than the typical cylindrical sampling tubes.[15] Because most sounders recovered enough material to examine microscopically, they satisfied Wallich's needs. But more famous microsopists than he had disagreed about whether the foraminifera found in bottom samples lived at great depths or fell there after death. The two scientists who examined Atlantic bottom sediment for Matthew Fontaine Maury in the 1850s, West Point professor Jacob Whitman Bailey and Professor Christian Ehrenberg of Berlin, argued this point vociferously. Wallich was one of the first microscopists to examine fresh, not preserved, marine sediments, and he felt as though he had the necessary experience to solve the problem of the extent of life in the sea.

Near the end of the *Bulldog* voyage, Wallich declared with satisfaction, "At the eleventh hour, and under circumstances most unfavourable for searching out its secrets, the deep has sent forth the long coveted message." In the 1,260-fathom sounding on October 15, thirteen living starfish (the genus of brittle stars) came up "convulsively embracing" the sounding line on the length of the rope just above the sounding apparatus. Immediately Wallich crowed, "As I have all along felt assured . . . the generally accepted notions as to the limits by which life is circumscribed in the sea must give way. A new source of study is laid bare to the Naturalist." The next day Wallich was even more self-congratulatory, "The more I reflected upon the discovery of the Star fishes, the more reason do I find to congratulate myself on being permitted to be the first to bring such a strikingly interesting fact to notice."[16]

The book Wallich published after the *Bulldog* expedition contained, in addition to a narrative of the voyage, a detailed exposition of Wallich's opinions on the bathymetrical limits of oceanic animal life. He tried to strike a balance between acknowledging John Ross's starfish and casting doubt on Ross's sounding in order to claim precedence for his own, very similar, accomplishment. He dismissed Ehrenberg's claim that zoophytes in deep-sea sediment samples had been alive when collected by declaring that Ehrenberg's observations did not justify his conclusions. Wallich presented complex, long-winded answers to the questions of how organisms could survive under the extreme conditions of great pressure and no light. Finally, he presented his starfish sounding as the final proof that life existed at great depths. Unfortunately for Wallich, though his work was noticed by naturalists, it was not widely accepted. The same criticism he leveled against Ross—that the starfish might have attached to the line on the way up—reappeared to detract from his achievement.[17]

The proof that Wallich wanted so desperately to provide came instead from a submarine telegraph cable. At a port stop during the *Bulldog* cruise, Wallich received a letter dated August 15 telling him that a cable had been raised from "no less than 1700 fathoms," pierced through with a small teredo, a burrowing mollusk related the the shipworm, that had been brought up alive. The letter concluded, "The fact may accordingly be considered settled for ever that organic beings live down at the bottom of the sea, at vast depths." While correctly interpreting the impact of the discovery, the letter had garbled some of the facts. The cable in question had been laid in 1857 between Sardinia and Bona, on the African coast. It failed in the summer of 1860, with the cause traced to a break at 1,200 fathoms. The engineer who raised it, Professor Fleeming Jenkin, found the cable encrusted with marine organisms, which he sent to George Allman, who held the natural-history professorship at Edinburgh University. Jenkin also turned his personal journal from the voyage over to the Edinburgh zoologist Charles Wyville Thomson for reference while Thomson wrote the historical portion of *The Depths of the Sea*. In addition, the French naturalist

Alphonse Milne-Edwards examined some of the organisms recovered and delivered a paper about them to the French Academy of Science.[18]

Thomson explained why naturalists accepted this discovery as proof of life at great depths while suspecting previous claims, including Wallich's. The fact that the cable was encrusted with marine animals proved that they had lived for a long time at the depth of the cable. This time the animals in question found their way into naturalists' hands, to be preserved, studied, identified, listed, and published. As scientific specimens, they became relevant to naturalists' work. Thomson interpreted Wallich's infamous starfish sounding this way: "The misfortune of these starfishes was that they did not go into the dredge; had they done so, they would at once have achieved immortality."[19]

By THE TIME reports arrived of the discovery of life on the Mediterranean floor, news about the scientific appeal of the sea had already spread from the remote northern coasts of Britain to the metropolis. Several active naturalist-dredgers had moved to London, most notably Forbes and Jeffreys. There they acquainted members of the scientific community with advances in marine zoology and its repercussions for other fields. A few Londoners delved into marine natural history in the context of seaside holidays. One of these, the physiologist and microscopist William Benjamin Carpenter, worked with Thomas Henry Huxley to seek government funding for marine research. They envisioned an extension of the Geological Survey to cover dredging along all of Britain's coastlines, a goal that they failed to achieve. Proponents of marine research soon concluded that beach holidays provided an insufficient basis for studying the sea.

Forbes and Jeffreys, two of the most devoted and enthusiastic dredgers, brought their passion for marine zoology to London. Both were members of the Royal Society and participated in other metropolitan scientific societies as well. Forbes moved to London in 1843 to take the chair of botany at King's College. To earn extra income, he also became curator of the Geological Society. His new responsibilities left him little time for research, so he eagerly accepted a job the following year as paleontologist for the Geological Survey. He held this position until he won the Regius Chair of Natural History at Edinburgh University in 1854, just before he died at the age of thirty-nine.

Although Forbes had less time to dredge during his years in London, he remained an enthusiastic proponent. His participation in metropolitan scientific circles introduced men of science to the hitherto hidden world beneath the waves. Just as Maury had attempted to create a new field of physical geography of the sea, so Forbes intended to promote marine natural history. His planned magnum opus, *The Natural History of the European Seas,* published posthumously in 1859, described the zones of habitation that Forbes identified to define geographic distribution of marine fauna and flora. Forbes's azoic theory was the first serious scientific attempt to formulate a statement about oceanic life.[20]

Soon after Forbes left London, Jeffreys arrived. Whereas Forbes piqued the interest of geologists, Jeffreys served as a repository for knowledge about marine zoology, particularly recent and fossil mollusks, and also as a contact with ocean scientists in other countries. Perhaps more important, he acted as the conduit for expertise about dredging technology and the techniques honed by provincial dredgers across Britain. Jeffreys arrived in London in 1856 to practice as a barrister and became involved with the Royal, Geological, and Linnean societies. When he retired from the law ten years later, he devoted himself to the study of European mollusks. Although he single-handedly orchestrated the British Association for the Advancement of Science (BAAS) dredging committees for a decade, Jeffreys was known by his contemporaries more for his important collection, called by Philip P. Carpenter, another well-known British mollusk specialist, "the finest European collection." As the keeper of such a collection, Jeffreys was a central figure who acted, in some ways, like an institution for British naturalists who specialized in the study of mollusks, malacology.[21]

Jeffreys also served as a consultant for London men of science on dredging and ocean science. He offered in 1866 to help arrange a berth for George Wallich to participate in a dredging excursion to either the north of China or the Straits of Magellan. When Charles Lyell wanted to know what naturalists expected to learn from deep-sea dredging, he turned to Jeffreys, who explained naturalists' search for the effect of temperature on marine life. Jeffreys's correspondence with foreign malacologists benefited the British scientific community. He encouraged foreign scientists to visit England, often hosting them or inviting them for dinner at his home. He himself traveled extensively in Europe, visiting museums and other important collections of terrestrial and marine mollusks. In 1871, Jeffreys visited the United States, where he spent time at the Museum of Comparative Zoology at Harvard with Louis Agassiz. He visited William Stimpson in Chicago just before the great fire that destroyed the invertebrate collections, fortunately taking with him a series of duplicates from Coast Survey dredgings in the Gulf Stream during 1868 and 1869. Stimpson's young friend William Healey Dall, who collected and studied marine invertebrates while doing fieldwork for the Coast Survey in the early 1870s, enjoyed exchanging letters with Jeffreys during Dall's brief winter returns to the San Francisco Coast Survey office between long and arduous field seasons in Aleutian Islands waters.[22]

As widespread as Jeffreys' scientific network was, it was limited to zoologists, indeed mostly to malacologists. In order for ocean science to gain a toehold in the metropolis, investigators of other parts of the natural world had to learn about the scientific promise of the sea. One of these, William Benjamin Carpenter, became one of the most active promoters of ocean science in Britain.

Like many dredgers, Carpenter first encountered the sea through travel and recreation. He earned a medical degree from Edinburgh University in 1839, at the height of the popularity of dredging excursions there. It is likely that Carpenter knew about the dredging activity. His first acquaintance with the open ocean came after a medical apprenticeship, when he traveled as companion to a patient to the West Indies, where he developed a passion for sailing and rowing. Although his early scientific research was in physiology, in the 1850s he moved into the fields of microscopy and marine zoology, focusing particularly on the larval development of marine fauna. In 1856, he published *The Microscope,* an extremely popular book aimed at conscripting amateur observers into the service of professional science. Influential in both England and the United States, the book continued through six editions, the last of which came out in 1881. Among its many enthusiastic readers were professional hydrographers interested in studying deep-sea bottom samples. Commander William Chimmo asked the Hydrographic Office in 1870 for a copy of Carpenter's book and a better microscope to study foraminifera in his spare time. Carpenter himself investigated recent and fossil foraminifera and published a monograph on the subject in 1862.[23]

Carpenter's serious pursuit of marine zoology started on a summer seaside holiday to Arran for his health. He admitted candidly, "I came here without much idea of doing more than amuse myself with Natural History." When beachcombing failed to yield interesting specimens, Carpenter questioned local residents and visiting naturalists, who advised him to dredge in Lamlash Bay. There he found "things rare or not common elsewhere." He took these "treasures" to the Glasgow British Association meeting, where they created quite a stir among an audience of knowledgeable dredgers. Convinced by this reaction that Lamlash offered unusually fertile ground, Carpenter returned to Arran after the meeting. A rich haul of laminaria, a type of kelp, rewarded him with many small, attached crinoids (sea lilies and their relatives), providing a superb opportunity to search for a larval stage. Finding none in so large a haul, he concluded that the larvae must be free swimming.[24]

Carpenter's enthusiasm for the zoological possibilities of Lamlash Bay inspired a scheme to establish a marine research base there. A zoological survey of the coast in conjunction with the Geological Survey had become a serious possibility for Huxley. When Huxley succeeded Forbes as lecturer at the Government School of Mines in 1844, he envisioned a full-scale survey of the marine fauna of British coasts. In Carpenter's letters to Huxley about the Lamlash Bay station, he referred to "the Coast Survey" as if he considered it already under way. In fact, Huxley thought of his 1854 dredging holiday at Tenby as the start of this plan. During his honeymoon there the following year, he dredged and studied specimens partly for his own research and partly for the Geological Survey.[25]

Huxley envisioned a project of much greater scale than his own summer investigations, although one that would mostly use existing resources. One government naturalist, namely himself, would collect and coordinate information from Coast Guard officers, fishermen, and local naturalists. He would visit parts of the coast as necessary, accompanied by a collector to take care of routine work. Huxley's scheme would essentially expand British Association efforts to make use of the working knowledge of mariners and fishermen. It would have transformed the study of marine fauna into an official government project, rather than the private, though national, enterprise it had become in the hands of British Association dredgers.

In 1855 Carpenter outlined his proposal to Huxley. He had already discussed his idea with people in the area, lining up housing for visiting scientists as well as skilled local labor for keeping and breeding marine animals. As important as the physical plant was the presence of willing local experts. Carpenter envisioned, in addition to traditional collecting activities, the establishment of "a great Vivarium in which we could get the animals *to breed.*" This was important for Carpenter's interest in identifying larval forms and studying development. Such an operation would require continuous supervision, but Carpenter had found just the person to run it. Glasgow in the 1850s was home to a small but active group of dredgers. Three years before Carpenter's Arran holiday, five BAAS dredgers worked off the east coast of Scotland. The summer Carpenter discovered Lamlash Bay, the BAAS convened a new dredging committee for Scotland's west coast, presumably inspired by the specimens Carpenter displayed at Glasgow. Two members of the previous dredging committee, Dr. Robert Greville and Professor John H. Balfour, were joined by a Mr.

Eyton and the Reverend Charles P. Miles. Carpenter proposed to Huxley that the Episcopal clergyman, Miles, be the overseer for the marine station and vivarium.[26]

The most immediate obstacle to Carpenter's scheme proved to be the lack of comfortable accommodation. Since Arran was a favorite resort of Glasgow residents, housing was difficult to find. The Duke of Hamilton, who owned nearly the whole island, forbade the construction of new homes and tore down old ones whenever possible. Given this problem, Carpenter was delighted to discover an appropriate house for a vacation marine station, one that offered easy access to Lamlash Bay. The house stood on Holy Island, which almost closed off Lamlash Bay and made it a calm and safe harbor for dredging. Much of the island's coast consisted of steeply rising rock, providing no convenient access to the sea. Near one end, though, was a moderate slope on which stood "a very good house," along with several outbuildings, a pier, a vegetable garden, and grazing ground. The house offered accommodations appropriate for a naturalist's family, a crucial requisite for the proposal's success. Equally important were good facilities for scientific work, most notably the wooden pier, which would allow a naturalist to "stand just over the very best part of the ground," ensuring good collecting opportunities even when dredging was impractical.[27]

Having worked out the practical details, Carpenter moved quickly to try to secure the property. His letter to Huxley laid out the financial considerations. To lease the entire island would cost £50 per year, although the current tenant was willing to keep his lease for the land and rent just the house for £25 or £30 per year. Given the zoological richness of the bay itself, plus the house's convenience as a staging point for operations elsewhere, Carpenter hoped to get the use of the property for a number of years. He asked Huxley to approach Sir Roderick Murchison, the new head of the Geological Survey of Great Britain, and present the location as an excellent one for a government Coast Survey station. In particular, since Murchison knew the Duke of Hamilton, who owned the property, Carpenter thought it possible that the duke might simply make the house available to naturalists.[28]

The Coast Survey's failure to materialize precluded the transformation of the house on Holy Island into a government marine station. Carpenter did not want to lose the opportunity entirely, though, so he made an offer for the house himself, for five years at £25 or £30 per year. During the next several summers, Carpenter collected and dredged in Lamlash Bay and the surrounding waters. Members of the BAAS dredging committee, including Balfour, Miles, and Greville, visited and based their operations there. An-

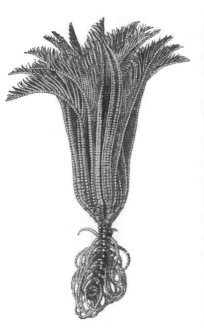

Two species of crinoids, or sea lilies, of the *Pentacrinus* genus, strikingly beautiful animals that were well known through fossils before being discovered in the deep sea. Naturalists, primed by Darwin's theory, wondered if the ocean's depths might prove a repository for living examples of species believed to be extinct. Although *Pentacrinus* fulfilled this expectation, few other examples surfaced in the *Challenger's* nets. (From Charles Wyville Thomson and John Murray, ed., *Report of the Scientific Results of the Exploring Voyage of the HMS "Challenger," 1873–76: Narrative of the Cruise,* vol. 1 [London: HMSO, 1885], 304.)

other visitor was the physicist William Thomson (later Lord Kelvin), who later contributed to the Atlantic cable project and promoted ocean science. George Barlee wrote to Joshua Alder in the summer of 1856 complaining about his bad luck in dredging that season until he heard from a local dredger about the Lamlash group. He traveled there to investigate, met Miles and Greville, and received a cordial invitation to stay for awhile. Barlee reported that seven naturalists shared the place, two residing at a time for two months each. They kept an aquarium and possessed all the boats and dredges they needed to do their work. That winter Alder wrote to Norman about the Holy Island arrangement, explaining its affiliation with the BAAS. The dredgers, of whom Alder knew by name only Miles, Carpenter, and Greville, received £10 from the Association in 1856 and £25 the following year.[29]

Although Carpenter never fully joined the ranks of BAAS dredgers, he did not forget about marine zoology after his summers at Lamlash. During the 1863 and 1864 vacations, he visited his friend Charles Wyville Thomson at Belfast to collaborate on a study of the larval stages of crinoids. Thomson seems to have visited Carpenter and the others at Holy Island.[30] From Thomson, Carpenter heard about the profound implications of recent discoveries from depths beyond those explored by most naturalist-dredgers.

Thomson was one of only a handful of university professors active in BAAS dredging. He joined his first BAAS dredging committee in 1852, two years after he began lecturing on botany at Aberdeen. In 1854, he moved to Belfast to take the geology professorship, but was named professor of zoology and botany there in 1860. In 1861 he provided drawings for Robert Patterson's Belfast dredging committee report. Two years later, he began his most protracted service, sitting for three years on the Belfast dredging committee at the height of its activity. After that, his appearances on committees were more scattered: the Shetland committee in 1862, the northwest Ireland committee in 1866, and the north Hebrides committee in 1880. Thomson met Jeffreys through the Belfast dredging committee when Jeffreys joined its members for a fortnight of collecting off Larne, a deep-water port north of Belfast on Ireland's northeast coast.[31]

Thomson's contact with a dredger outside the BAAS community spurred a renewed search for life in the deepest ocean basins. In 1866, Georg Ossian Sars, working in his capacity as superintendent of Norwegian fisheries, dredged a stalked crinoid near Lofoten from a depth of 300 fathoms. While crinoids grew in coastal areas, none resembled the many stalked crinoids preserved in the fossil record. This discovery, coming not long after the publication of Darwin's *Origin of Species,* drew attention to the ocean's depths as a possible storehouse of "living fossils." Thomson traveled to visit Sars and his father and teacher, Michael Sars, shortly after Georg captured this prize specimen. The elder Sars was a Norwegian clergyman who, in 1854, had been appointed to the professorship of zoology at the University of Christiana (now Oslo). British marine zoologists recognized and appreciated his contributions to the knowledge of marine fauna from "the wildest part of Norway." Thomson and other zoologists therefore took notice of Sars's reports of abundant and varied animal life at depths down to 450 fathoms.[32]

Sars used the naturalists' preferred tool, the dredge, and gave Thomson crinoid specimens that he showed to Carpenter in Belfast in early 1868. Impressed by the types and varieties of organisms found at that depth, they argued that Sars's collections "place it beyond a doubt that animal life is abundant in the ocean at depths varying from 200 fathoms (1200 feet) to 300 fathoms (1800 feet)." To explore this remote region Carpenter and Thomson hatched a scheme for obtaining Admiralty assistance for deep-sea dredging.[33]

NATURALISTS FACING difficulties reaching the depths they yearned to explore grew dissatisfied with rowboats and yachts. Early on, dredgers

Naturalists welcomed navy sailors' help in hauling up dredges from deep water and sorting their contents. Steam vessels had allowed dredgers to travel to and from collecting sites more quickly than sailing yachts, but mechanical assistance for hauling in collecting gear did not become available until 1867. (From Charles Wyville Thomson and John Murray, ed., *Report of the Scientific Results of the Exploring Voyage of the HMS "Challenger," 1873–76: Narrative of the Cruise*, vol. 1 [London: HMSO, 1885], 166; courtesy of Dr. David C. Bossard.)

solved this problem by hiring steam yachts. The increasing number of landlubbers discovering marine zoology appreciated their ability to set out for an afternoon's work even if the weather appeared chancy, which it often did in the northern waters, because they knew they could return to port quickly. Jeffreys hired a steamer one summer in Shetland, and BAAS dredging parties sometimes did too. In 1865 dredgers planned to rent a tug steamer in Aberdeen for a week of work in July.[34] Employing working vessels with professional sailors relieved dredgers of the need to maintain a yacht and supervise a hired crew. Yachtsmen who dabbled in science enjoyed the trappings of the sport, but professional scientists increasingly preferred to hire boats and crews for only the time they needed them. Dredging from rented steamers had drawbacks, though, notably the expense. And although steamers made transportation to and from dredging sites more reliable, they did not alleviate the considerable labor of hauling the dredge in by hand.

Naval assistance began to seem an attractive solution to all these problems. The application of steam power to hydrographic ships in the 1850s made large vessels maneuverable enough for conducting sounding and dredging from their decks instead of from small boats. Mechanical assistance with hauling in dredging and sounding lines came only later; even in 1867, Captain J. E. Davis recommended that crew members be provided with rawhide hand protectors to use when hauling in small lines with heavy weights. Consider the effort involved in deep-water sounding: An 1821 sounding of 1,000 fathoms took 22 minutes for the line to reach bottom and another hour and 20 minutes for 100 men to haul it in. In 1868, it took 62 men 2 hours to haul in 4,200 fathoms of sounding line from a depth that was later determined to be 2,600 fathoms. That same year, Davis, in command of HMS *Cordelia,* used a steam winch for hauling in weighted sounding lines. Although the crew members who were freed of the onerous task were certainly grateful, the winch still took 5 hours to bring back 1,850 fathoms of line.[35]

Thomson and Carpenter believed that the Admiralty might assist in deep-ocean exploration in part because of the enthusiasm with which the Hydrographic Office engaged in deep-sea sounding during the 1860s. Hydrographer John Washington oversaw the Royal Navy's first forays into great depths, but it was his successor, George M. Richards, who presided over the Hydrographic Office during the years when British hydrographers and naturalists banded together to study the deep sea. The Hydrographic Office increased in size during the 1860s, reflecting the expansion of commerce in all parts of the world. By the time Richards took office in 1863, submarine telegraph entrepreneurs and supporters had rallied to prepare for the second set of attempts to lay an Atlantic submarine telegraph cable. The prospect of an Atlantic cable transformed the ocean floor into a politically important area and brought it to the attention of Admiralty officials, politicians, and investors, as well as the general public. As the strategic value of submarine cables became increasingly apparent, this impulse to keep naval resources at the ready translated into a mandate to secure the deep-sea floor for British cables. Under Richards's tenure as Hydrographer, the Royal Navy routinely conducted the surveys for all proposed British cables in every ocean. It was not until the 1870s that cable companies acquired their own survey vessels.[36]

During sea-floor surveying voyages, hydrographers created the technology that enabled dredgers to reach the vast depths they sought. As American hydrographers working under Maury did in the 1850s, those sailing for the British Hydrographic Office continued to experiment with sounding technology. Many improvements derived from modest, piecemeal changes to sounding devices, such as those made when Commander William Chimmo undertook trials of a sounding machine invented by Lieutenant Fitzgerald of HMS *Cordelia* during an 1866 deep-sea survey in the Bay of Fundy.[37] Other innovations involved sounding practices. Captain Richard Hoskyn, of HMS *Porcupine,* experimented with several techniques, settling upon the simultaneous use of a light line without a sampling device for measuring depth and a stronger line for recovering bottom sediments and taking temperature measurements. That way, if the strain of hauling in the sampling device broke its line, Hoskyn at least had a depth measurement to record.

Hoskyn continued in the tradition of Joseph Dayman, who conducted the first British trans-Atlantic survey, and other deep-sea hydrographic pioneers by submitting a detailed report of his sounding efforts, which the Hydrographic Office published as a pamphlet and distributed to other hydrographers. The experience of Hoskyn and other British and American hydrographers was summarized in Davis's 1867 pamphlet, "Notes on Deep-Sea Sounding." In it, Davis specified an old standby, Brooke's rod, as the best available sounder.[38]

One voyage in 1868 resulted in major improvements to deep-sea sounding equipment that helped hydrographers solve the problem of dredging in deep water. Captain Peter F. Shortland's 1868 cruise in HMS *Hydra* resulted not only in a new sounding device that carried the ship's name but also new techniques and rigging. During the cruise Shortland's crew designed a sounding device fitted with a core sampler, based partly on the design of Brooke's sounder. The blacksmith, Gibbs, and two unnamed crew members were paid £1 each for the idea of a spring that threw off the sounding weights after the weights drove the corer deep into the sediment. Stacked detaching weights allowed the use of more sinkers in deeper water.[39]

In 1868, deep-sea sounding remained an uncertain exercise. Determining the moment the weight struck bottom still involved a combination of observation and guesswork. Captain Edward Calver said of deep-sea work, "Respecting sounding, observations of a critical character and sufficiently numerous have not been made, so as to admit of the operation being reduced to the precision of a formula." After Shortland's crew hauled up a few empty sounding devices, Shortland learned to let the line run out well beyond where he judged the weight would strike bottom. Soon he realized that the lengthened time intervals after hitting bottom would be more obvious if, instead of running free, the sounding line was resisted by a force equal to the weight in the water of a length of line of the expected depth. He ordered the construction of a sounding reel larger and stronger than the ones the *Hydra* or other sounding vessels had been issued. He then fitted a brake to the reel to control the velocity of its rotation.[40]

Applying resistance to the sounding line increased the chances that a

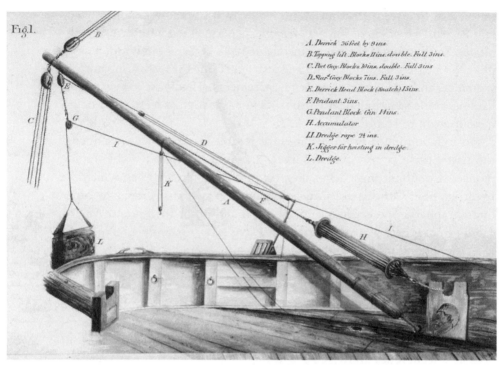

Fig.1.

A. Derrick 36 feet by 9 ins.
B. Topping lift. Blocks 11ins. double. Fall 3 ins.
C. Port Guy. Blocks 10ins. double. Fall 3 ins.
D. Star.ᵈ Guy. Blocks 7ins. Fall 3 ins.
E. Derrick Head Block (Snatch) 15 ins.
F. Pendant 3 ins.
G. Pendant Block. 6in. 14 ins.
H. Accumulator
I.I. Dredge rope 2¼ ins.
K. Jigger for hoisting in dredge.
L. Dredge.

The bow derrick of HMS *Porcupine* as rigged for deep-sea dredging, with accumulators to protect the line from breaking when the ship tossed while the dredge was deployed. Captain Peter F. Shortland of HMS *Hydra* pioneered this rigging for deep-sea sounding and Captain Edward K. Calver applied it to dredging. (United Kingdom Hydrographic Office, Original Document Series 394, fig. 1; © British Crown Copyright 1994; reproduced by permission of the Controller of Her Majesty's Stationery Office and the UK Hydrographic Office [www.ukho.gov.uk].)

ship's rolling and pitching would jerk the sounding line beyond its breaking point. To prevent this, Shortland added a device called an accumulator to the sounding tackle, in order to ease the strain on the line. The accumulator consisted of many vulcanized india-rubber springs fastened together at both ends, but kept from tangling each other by being run through holes punched around the edges of two round wooden disks. Rope between the two disks allowed the accumulators to stretch two or three times their length, but not past their breaking point. Not only did the accumulator keep the sounding line from breaking, but it served as a visual indicator of the amount of strain on the line. Shortland also hit upon the idea of attaching a tape measure to the sounding derrick to keep track of how far the accumulator stretched. In practice, he found he could estimate the strain simply by glancing at the angle of the sounding line as it passed through the block attached to the accumulator, which normally varied between 90 and 120 degrees.[41]

The Hydrographic Department praised Shortland's invention of the *Hydra* sounder and the efficacy of his "mode of obtaining deep sea soundings." His rigging and technique profoundly altered deep-sea sounding because they required the installation on hydrographic vessels of bulky, permanent machinery. But such reels could easily be attached to steam engines, so Shortland's rigging promoted the application of steam to what had been the manual work of hauling in sounding devices. In 1869 the *Porcupine* crew enjoyed a shorter and easier 2,435-fathom sounding using a steam engine. The sounder descended for 33 minutes, then the crew watched, no doubt with relief, as the steam engine heaved it up for 2 hours. But although this new equipment meant less work for the sailors, it required careful supervision by officers and engineers.[42]

In light of the increased scale of operations necessary to collect from deep water, would-be ocean scientists turned to the Admiralty. Carpenter insisted that dredging in deep water would require "a vessel of considerable size . . . with a trained crew such as is only to be found in the Government Service." He and his fellow advocates continued to embrace the moral and physical benefits of strenuous outdoor exertion in the name of science. Most of them, though, were professionals, not amateurs. To these men, collecting excursions represented a legitimate part of their scientific work, not a holiday activity parlayed into an achievement of scientific value. Thomson even insisted that deep-sea science "is scarcely compatible with pleasure-seeking."[43]

In May 1868, Thomson wrote to Carpenter in his capacity as a Royal Society Council member, proposing that the Society petition the Admiralty to organize a deep-sea sounding expedition. Thomson's ambitious plans called for dredging at a depth of 1,000 fathoms. He expressed confidence that "a couple of miles of stout manilla [*sic*] rope" would do the trick. He expected that the procedure would be quite straightforward for anyone accustomed to dredging, but felt that the scale of deep-water operations necessitated navy support. Thomson viewed the project as not much larger in scope than the work of BAAS dredging parties that hired steam vessels. Knowing that the Admiralty frequently made small surveying ships available to assist cable companies, Thomson thought that one of these might as easily be spared to carry out scientific research.[44]

Less than three weeks after receiving this letter, Carpenter wrote to

General Edward Sabine, then president of the Royal Society, enclosing Thomson's letter and outlining his and Thomson's plan. Carpenter stressed the scientific importance of studying the deep ocean. He noted that recent discoveries in the deep sea, especially by Sars, had aroused interest in many branches of science, including zoology, paleontology, and geology. He asserted that private individuals, even with grants from scientific societies, could not successfully undertake deep-sea collecting work. He paused to explain that dredging at great depths required large vessels and, most critically, trained crews. Carpenter had thought more specifically than Thomson about the practical elements of their request. He curtailed the geographic range and depth of Thomson's proposal in the interest of avoiding the extra expense of a coaling vessel. For the same reason he expressed a preference for a boat capable of making way under both canvas and steam. He also requested a grant from the Royal Society for preserving alcohol, jars for storing specimens, and other scientific equipment.[45]

Hydrographer Richards responded warmly to informal inquiries as well as to the official overture made by the Royal Society in June. The Admiralty offered HMS *Lightning*, a paddle-sloop with the "somewhat doubtful title to respect of being perhaps the very oldest paddle-steamer in her Majesty's navy." Carpenter and Thomson departed from Oban, Scotland, in June 1868 and spent several weeks aboard *Lightning*, dredging and taking temperature measurements in waters to the north and west. Although Thomson had to leave the ship in early September, Carpenter remained aboard for six weeks. If the scientists had anticipated being assigned this "cranky little vessel," they might have started with a less ambitious proposal. The ship leaked badly, a serious problem given the deplorable weather they encountered. Consequently, in Thomson's understated words, "we had not good times in the 'Lighting.'" The discomfort and danger rendered all the more marvelous the accomplishment of dredging in 650 fathoms, almost twice as deep as previous dredgers had ever reached. Even at this depth, they found abundant and varied animal life. They returned with enough specimens to convince any naturalist who still doubted the existence of life at extreme temperatures and pressures.[46]

But despite this achievement, the work had barely begun. *Lightning*'s few scattered deep dredgings whetted the appetite of naturalist-dredgers, so long limited to the first few fathoms of the sea. Mere days after disembarking, Carpenter assured Alfred Merle Norman that their deepest sounding that summer "was accomplished with such facility that I am satisfied . . . we could without difficulty have worked our dredge at 1000 fathoms."[47]

Lightning's dredging success derived in part from the new sounding technology championed by Captain Shortland. Captain Daniel May found the india-rubber accumulators "very useful and [was] sure they several times saved dredge and rope." The reel, accumulator, and other components proved as effective for deep-sea dredging as for sounding, and the application of steam power made dredging much easier. *Lightning* was fitted with a donkey engine for its "special work." Captain May warned, however, that working in water deeper than 1,000 fathoms would require a stronger engine than that carried aboard *Lightning*.[48]

On the basis of these limited but encouraging results, the Royal Society convened the Committee on Marine Researches, consisting of Carpenter, Richards, and John Gwyn Jeffreys, in addition to its president and officers. The committee report stressed the value of submarine research for physical, biological, and geological science. It presented the deep-ocean floor below 500 fathoms as "a vast field for research of which the systematic exploration can scarcely fail to yield results of the highest importance." Before the committee made specific recommendations for a cruise to follow up the work of the *Lightning,* it issued a general call to the government to undertake scientific study of the deep-ocean floor as "one of the special duties of the British Navy."[49]

The Admiralty responded within a month to the Royal Society request, assigning HMS *Porcupine* for four months of special scientific work. Whereas the *Lightning* voyage had been inspired by the search for living fossils, *Porcupine* organizers planned from the start a joint zoological–physical science investigation. The *Porcupine* cruise had two primary objectives: "to carry down the [physical and biological] survey to depths much greater than have yet been explored by the Dredge" and to investigate the cold and warm areas that *Lightning*'s temperature soundings had revealed in the Faroe-Shetland Channel.[50]

Although the *Lightning* voyage had been primarily devoted to dredging, Hydrographer Richards had used the opportunity to test deep-sea thermometers. Unexpected temperature readings from great depths intrigued Carpenter, who began to devote his energy to developing a theory of general oceanic circulation. Never entirely a naturalist-dredger, Carpenter became more and more absorbed by the physics and chemistry of the sea. His interest, and indeed the strong interest of many Royal Society members and Admiralty officers, in oceanic physical phenomena was fanned by the submarine telegraph enterprise and helped win government patronage for deep-sea study. Naturalists such as Jeffreys and Thomson were only too happy to recruit the powerful influence of physical science to the cause of ocean exploration.[51]

But initially even the cachet of physical science was eclipsed by the excitement of exploring the new territory of the ocean's depths. HMS *Porcupine* set sail in May 1869 for the first leg of its summer work with Jeffreys as scientific director. Within the first weeks of the voyage, *Porcupine*'s dredge returned full of Atlantic ooze from "no less than 808 fathoms," thanks in part to the double-cylinder, high-pressure, auxiliary engine of six horsepower. Proudly chronicling this moment, Jeffreys informed Edward Sabine, "This is the greatest depth ever reached anywhere with the dredge." Soon after, the crew conducted a dredge haul from 1,476 fathoms that yielded mollusca, a stalk-eyed crustacean with unusually large eyes, and a fine specimen of *Holothuria tremula*, a deep-water sea cucumber.[52]

Jeffreys's presence on *Porcupine* helped to ensure the voyage's success. During his Shetland yachting voyages, Jeffreys had developed techniques for dredging in deeper water than any other naturalist had sampled. He also contributed to the enterprise through the skill of the paid assistants he had trained, whom he recommended for the cruise. William Laughrin, a coast guard man and an associate of the Linnean Society, helped with the dredging itself and also the sifting of mud. B. S. Dodd worked on picking out, cleaning, and storing specimens. Both of these men remained with the vessel all summer, providing valuable continuity through the second and third legs of the voyage headed by Carpenter and Thomson.[53]

However valuable Jeffreys's dredging experience, he himself acknowledged that the achievements of the *Porcupine* cruise, "could not have been attained without the invaluable aid of Capt. Calver, who has brought all his scientific skill & experience to bear on the expedition and has not spared any means or trouble to ensure success." Edward Calver had begun his naval career at the age of fifteen and thereafter seized every opportunity to advance his skill in surveying, for which he showed an early aptitude. After eight years of service, he was appointed to the Hydrographic Office as an assistant surveyor. For the next decade, he conducted offshore surveys, wrote reports on harbors, and compiled sailing directions. In 1863, he took command of the *Porcupine* from Hoskyn, who had just finished his cable survey off the Irish coast. Calver dabbled in natural history enough to understand the objectives of the scientists with whom he worked. He attended BAAS meetings as early as 1853 and became an active supporter of the museum in his home town of Sunderland, on the North Sea in northern England, eventually donating to it a large collection of specimens from the *Porcupine*'s cruises. After retirement, he was elected a Fellow of the Royal Society.[54]

Porcupine was a small but seaworthy and steady vessel, well suited for surveying work. Its crew members were Shetland watermen who had, by

1868, already spent several summers under Calver's command. The officers had likewise been engaged in surveying work from *Porcupine* for several years. The crew worked well together and quickly learned to apply their sounding skills to deep-water dredging. Calver emphasized the importance of their experience, claiming that "practical skill is the principal requisite" for such work. As with deep-sea sounding, *Porcupine*'s dredgers experimented at sea with combinations of equipment and techniques that allowed them to reach bottom. One day early in the first cruise, the two-inch dredge rope anchored the ship suddenly. The strong line held fast, causing the ship to swing around the dredge line as an anchored ship swings around the chain cable holding its anchor. The officers surmised that the dredge had caught on a rock and confirmed their guess when the dredge finally returned with a piece of solid limestone torn from the bottom. Calver and his crew learned by doing, and applied their accumulating experience to conquering the depths.[55]

Porcupine had been supplied with a version of the typical northern British dredge—one based on Ball's model, although larger. Calver reported to Hydrographer Richards that the dredges were "far too light for the special work we are about." He expressed astonishment that Captain May had not noticed this during the *Lightning* voyage. Perhaps the absence of Jeffreys from *Lightning* accounts for this oversight, since he had long before rejected light dredges for deep-water work. Calver estimated that a dredge of 300 pounds, rather than the 100-pound ones supplied, would be necessary to dredge in the 1,000-fathom range. To solve the problem while not having access to the resources of a Royal Navy dockyard, he improvised by paying a blacksmith at Galway to "to add to the weights of one of our small dredges and to make certain improvements in form which experience has suggested."[56]

As in commercial dredging, it had become customary for naturalist-dredgers to allow a scope of line four times the depth of the water. Maintaining this proportion proved impossible for the depths dredged by the *Porcupine* because "no deck could conveniently contain the necessary amount of rope." So, as *Porcupine*'s dredgers reached deeper into the sea, Calver began to add auxiliary weights to the dredge line above the dredge at a distance of one quarter of the depth of the water. This technique helped the line sink almost as fast as the dredge and allowed Calver to dredge with a length of line only twice as long as the depth. For the deepest dredge haul the rope more nearly approached the depth. Even so, Calver's weighted line did sink, after which he allowed the ship to drift for two hours, "to ensure scraping along the bottom."[57] The success of this haul not only proved the effectiveness of the donkey engine–accumulator rigging but also demonstrated the plausibility of a method that departed from traditional dredging techniques.

Another dredging innovation of Calver's seemed almost miraculous to naturalists. Dredging most often yielded many pounds of dense, sticky mud, because the dredges dug into the seabed and became so clogged with fine-grained sediment that everything else was pushed aside. On stony surfaces dredges often returned empty in places where "it was fair to suppose there was an abundance of animal life." To combat these problems Calver had the idea of sending down, attached to the dredges, "swabs," the frayed and knotted tangles of rope used to wash the deck. He explained, "The idea of the Tangles was suggested from noticing two or three small crustaceans entangled by the frayed-out end of the dredge rope, and after a failure or two in its operation, a plan was devised which succeeded most perfectly." An attempt to attach the tangle to arms projecting out from the rope above the dredge failed because the rope was often lifted off the bottom. But when the *Porcupine* dredgers fastened the tangles to a long iron bar transverse to the bottom of the dredge bag, the results were "uniformly successful." Thomson declared, "We now regard the 'hempen tangles' as an essential adjunct to the dredge, nearly as important as the dredge itself, and usually much more conspicuous in its results." Although the "ingenious device" multiplied the the number of the naturalists' deep-sea prizes, it sometimes made a "sad mess" of the specimens, whose mangled remains woeful scientists clipped out of the swabs with nail scissors. They consoled themselves with the few whole creatures recovered, and reflected "had we not used this somewhat ruthless means of capture, the mutilated specimens would have remained unknown to us at the bottom of the sea."[58]

The success of the first leg of the voyage under Jeffreys's direction "so far realized [their] anticipations" that *Porcupine* organizers dramatically revised plans for the second leg. Instead of exploring the area west of the Outer Hebrides, they determined to attack head on the question of the bathymetrical limits of life. Since the deepest confirmed soundings did not extend much beyond 3,000 fathoms, they felt that they could "virtually solve for all the depths of the ocean" the question of the existence of life at great depths. Thomson, scientific director of the second cruise, consulted with Calver and found him eager to attempt dredging in the deepest area within their reach. With the Hydrographer's blessing, *Porcupine* headed for a point 250 miles west of Ushant, France, where the chart indicated 2,500 fathoms. On July 22, they completed a "perfectly successful" haul in 2,435 fathoms in a spot just past Ireland to the west, as far south as Ushant. They retrieved the now familiar deep-sea ooze and a wide variety of animals. Another attempt at the same depth brought up an empty dredge when the rope fouled around the bag, but a second successful haul at 2,090 fathoms confirmed the first.[59]

By the time the *Porcupine* proceeded to Stornoway, Scotland, to pick up Carpenter for the final leg of the voyage, the members of the expedition staff were satisfied that they had more than fulfilled their expectations. But, as Thomson reported, "these enormously deep dredgings could not be continued. Each operation required too much time, and the strain was too great, both upon the tackle and upon the nervous systems of all concerned." Indeed, the first "abyssal" dredging took an hour and 5 minutes for the whole of 3,000 fathoms of rope to run out, then 4 hours and 10 minutes to haul in, even with the steam engine. They therefore reverted to their original plans for the third segment of the voyage. At Stornoway, one of Carpenter's son's, P. Herbert Carpenter, joined the ship. He had sailed on the *Lightning* as an apprentice to John Hunter, the chemist for that voyage. On the *Porcupine,* Herbert was prepared to undertake by himself the detailed chemical and temperature analysis the scientists hoped would allow them to characterize the warm and cold regions they had discovered on the *Lightning.* Although the *Porcupine* scientists continued dredging throughout this leg, they confined their zoological collecting to the 400- to 700-fathom range.[60]

Zoologists were pleased with the *Porcupine* cruises because almost every specimen collected "would be accounted an important acquisition to Museums already most complete." Most scientists could share the zoologists' satisfaction upon learning of the abundant and varied creatures found at every depth. When Carpenter presented the results of the 1869 *Porcupine* voyage to the Royal Institution, he addressed topics of interest not only to naturalists but to a wide range of scientists. He spoke about the temperature discoveries at great depths and related the dredgers' conclusion that temperature exerted a greater influence on oceanic faunal distribution than pressure. He asserted the probability of finding many more living fossils. He related Thomson's idea about the continuity of the chalk, that the deepest portions of the Atlantic had been under water since the Cretaceous period, accumulating deposits continuously since than time. He did not forget that measure of success beloved by naturalists—the number of new species, and those species new to British waters, that were found during the course of the expedition.[61]

When Carpenter wrote a letter in March 1870 to request Royal Society backing for another cruise, he received an almost immediate response. In short order, a Royal Society committee convened and dispatched a letter to the Admiralty. Following the precedents of 1868 and 1869, the Admiralty replied promptly with the news of the *Porcupine*'s availability the following summer. Unlike the previous two summers' work, the 1870 cruise would devote half its time to studying Mediterranean undercurrents and the mixing of Atlantic and Mediterranean waters. As before, Jeffreys took scientific charge of the first leg, a two-month cruise from Falmouth to Gibraltar that took place while Thomson and Carpenter were still occupied with university duties. Plans to divide the work into three cruises changed when Thomson became ill, and Carpenter supervised all the work during the final two legs, from Gibraltar until Malta.[62]

By this time, Carpenter's interest had focused on oceanic circulation. He left the zoological research to his colleagues, although he still supervised dredgings along the Mediterranean cruise track. Though Carpenter had argued for the Mediterranean investigation in terms of zoological, geological, and physical science dividends, he especially stressed the temperature sounding work. He pointed out the need for new measurements because those made in the 1850s had been rendered worthless by new understandings of the behavior of thermometers under pressure. As for deep-sea dredging, Carpenter again concentrated on exploring depths from 400 to 800 fathoms, from which, "as experience has shown us, the most interesting collections are to be made." Indeed, most of the *Porcupine*'s collections from that summer came from that depth range.[63]

Advocates of deep-ocean exploration did not protest the *Porcupine*'s goals because they already had in mind a much larger-scale attempt to reveal the secrets of the sea. Thomson called the ocean floor "the land of promise," virtually the only place left on earth where scientists could collect "endless novelties of extraordinary interest."[64] He and others envisioned a major expedition to explore the floor of every ocean.

EVEN AS PLANS went forward for the 1870 *Porcupine* cruises, its organizers laid the groundwork for a more ambitious project, a circumnavigation of several years' duration. As the scheme took shape, Carpenter planned an appeal to the presidents of the British Association and the Chemical, Geographical, Geological, Linnean, and Zoological societies for their support. Starting with the second *Porcupine* leg that summer, Carpenter began to emphasize the physical sciences in his proposals to the Admiralty, revealing his own growing interest in oceanic circulation and reflecting his awareness that proposals including physics in addition to zoology and geology met with greater enthusiasm. Indeed, the Admiralty's response to the Royal Society's proposed circumnavigation expedition overtly emphasized physical science over the "other subjects therein named." The Admiralty said that "science and navigation" would profit from such a comprehensive examination of the "physical conditions of the deep sea."[65]

The organizers of what became the *Challenger* expedition drew on precedents established by the *Lightning* and *Porcupine* voyages, starting with their simultaneous use of official and informal social channels to test the waters. Carpenter in particular felt strongly that "personal influence is often more valuable than official representation." Thus by the time Carpenter had broached the idea of the *Lightning* and *Porcupine* cruises, he usually had in mind a particular ship that he knew was available. Likewise, for the circumnavigation expedition he first informally contacted Hydrographer Richards and other influential people. By the time he committed his plan to paper in a November 3, 1869, letter to Sabine, he had "already broken ground in the matter with Mr. Lowe and Mr. Childers." Robert Lowe was the Chancellor of the Exchequer and H. C. E. Childers was First Lord of the Admiralty. Because of his preliminary work, Carpenter could assure the Royal Society that its proposal would be welcomed by the Hydrographer and "favourably considered" by the Admiralty.[66]

The use of personal channels to organize expeditions characterized other voyages of ocean science that used British government resources. After the 1870 *Porcupine* cruises, neither that vessel nor the less appealing *Lightning* was available to continue the research begun during the summers of 1868–1870. Thus no plans were made in the early months of 1871 for another small-scale cruise. Instead, Carpenter and others concentrated on bigger plans for the circumnavigation. In early June, however, Hydrographer Richards informed Carpenter of an unexpected opportunity to extend his work on the Gibraltar currents. Starting in October 1871, Captain George S. Nares would be engaged in surveying in the Mediterranean. Before then, his ship, HMS *Shearwater,* would be available for temperature sounding work, if Carpenter was interested.[67] Hoping to demonstrate the presence of an outflowing undercurrent in the Straits of Gibraltar, which would lend support to his theory of density-driven rather than wind-driven deep-ocean currents, Carpenter joined the *Shearwater* in August. This cruise proved significant for the organization of the circumnavigation expedition because Nares later left *Shearwater* to command *Challenger,* bringing with him two of his lieutenants. It also demonstrated Admiralty willingness to take an active role in planning physical studies of the ocean.

In November 1871, the Royal Society convened the Circumnavigation Dredging Committee, choosing that name despite the expanded goals that included physical science. The group consisted of Hydrographer Richards, Royal Society officers, and advocates of ocean science, including committed dredgers such as Carpenter, Jeffreys, and Thomson, as well as the well-known scientists Thomas Henry Huxley, Joseph Hooker, and William Thomson. With Admiralty and Royal Society support in place, the next step was to convince the government to finance a major ocean science expedition. Carpenter enlisted influential men of science to testify to Cabinet members about the scientific importance of their proposed "Circumnavigation expedition." Carpenter suggested that Sabine invite Lowe and Childers to a Royal Society meeting while the Cabinet was in session. Royal Society Fellows who were also Cabinet members would be requested to attend. Carpenter planned to ask representatives of various scientific fields to "express their sense of the value of [deep-sea investigation] and their desire for its extension."[68]

Carpenter's list of expert witnesses included some of the most eminent and active members of the London scientific community, several of whom served on the Royal Society's circumnavigation committee. He had in mind Sir John Herschel for physical geography, Sir Charles Lyell for geology, Thomas Henry Huxley for zoology, George G. Stokes or John Tyndall for physics, and William Allen Miller for chemistry. Sabine, he planned, would "sum up the whole." Carpenter intended to impress the government officials with the testimony of these well-known scientific figures that studying the ocean's depths would yield great scientific dividends and maintain Britain's supremacy over the sea.[69] In doing so, he was hardly asking these men to lend their name to an enterprise with which they were unfamiliar or indifferent.

Indeed, they and other London scientists were well persuaded of the scientific importance of the deep sea for physical and chemical science as well as zoology and natural history. Herschel had corresponded with Carpenter about the issue of deep-sea temperature, both the practical problems of measuring it and the theoretical implications for oceanic circulation. He also communicated with Matthew Fontaine Maury, sending him an article about physical geography and receiving in return proofs of Maury's most recent edition of *Sailing Directions* describing his deep-sea investigations. Herschel discussed Hydrographic Office results of deep-sea soundings and temperature measurements with Lyell, with reference to submarine telegraphy as well as circulation theories and the debate about life at great depths. Herschel owed his awareness of advances in ocean science in part to his position as editor of the comprehensive field guide *A Manual of Scientific Inquiry*. This book had sections on hydrography, geography, geology, and zoology, each written by an expert in the field.[70]

Lyell, like Herschel, had corresponded with Carpenter about ocean circulation as well as about subsidence, because Thomson's theory of the continuity of the chalk opposed Lyell's belief that the ocean floor had experienced repeated upheavals. Lyell also corresponded with Jeffreys about fossil shells, chalk formation, and the effect of temperature on marine life. Lyell was sufficiently interested in deep-sea research to ask Hydrographer Richards about plans for the publication of the 1869 *Porcupine* voyage results. Like Lyell's, Tyndall's research intersected with debates about ocean circulation.[71]

Carpenter kept Stokes apprised of the emerging plans for a circumnavigation voyage, emphasizing the responsibility, as he saw it, for the Royal Society to take the initiative in organizing it. Stokes, who was secretary of the Royal Society, not only sat on the committee that lobbied for and planned the *Challenger* expedition, but also had been a member of the earlier Royal Society dredging committee that organized the *Lightning* and *Porcupine* cruises. William A. Miller, treasurer of the Royal Society and a member of the committee that arranged the 1870 *Porcupine* voyage, suggested an improved deep-sea thermometer.[72]

Of the eminent scientists that Carpenter proposed to call to witness to the scientific and national importance of a circumnavigation expedition, none had as much knowledge or direct experience of the sea as Thomas Henry Huxley. An outspoken member of the London scientific community, especially in his self-assigned role as Darwin's spokesman, Huxley was well known as an expert on microscopic marine fauna and marine invertebrates. While aboard HMS *Rattlesnake* in the official capacity of surgeon, Huxley had employed a microscope and collecting net to study pelagic invertebrates and microscopic marine forms. Although he shifted his attention to vertebrates after 1854, he spent his honeymoon the following year dredging, and he continued to examine deep-sea sediment samples from hydrographic cruises long after that time. These samples were deposited in his care at the Museum of Practical Geology. In 1868, he reexamined a series of bottom samples and discovered a protoplasmic primitive organism that he believed to be a precursor of higher life forms. He christened the creature *Bathybius haekelii*, after Ernst Haeckel, who only two years before had established the third biological kingdom, the "Protista." *Bathybius*'s arrival, awaited by geologists, cytologists, protozoologists, and evolutionists, whose lines of research had converged in the 1860s, fulfilled

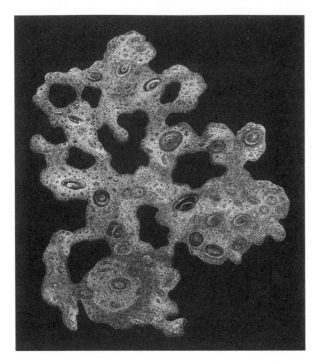

Bathybius haekelii, whose identification by Thomas Henry Huxley as a proto-plasmic link between life and nonlife spurred a global search for it by the *Challenger* scientific party before this mysterious creature was debunked, found to be a chemical precipitate formed by the reaction of seawater with preserving alcohol. (From Charles Wyville Thomson, *The Depths of the Sea* [London: Macmillan, , 1873], 412.)

many scientific expectations. It provided a subcellular, elemental unit of life that served as the bridge between life and inorganic matter and promised to explicate Darwinian theory.[73]

Because of Huxley's naval experience, his opinion about scientific matters deeply impressed naval officers and officials. Through his position as government inspector of fisheries, Huxley was often consulted by officials on marine matters. Commander Joseph Dayman referred to Huxley's research on the composition of the ocean bottom during his testimony before the commission that investigated the failed 1858 submarine telegraph cable. Later, Lord George C. Campbell, an officer on HMS *Challenger,* invoked Huxley's name to help him bear the extreme frustration, boredom, and impatience of the final month of the expedition, assuring himself, "Has not Huxley said that our work in the Atlantic alone more than repaid all the bother and expense of the outfit?" In addition to his general support of ocean science, Huxley materially aided the expedition by taking over Thomson's courses in Edinburgh for the final two years of the voyage and later by writing a volume of the *Challenger* reports.[74]

Other well-known scientists supported ocean science in a general way, especially those who, like Huxley, had direct experience of seagoing and studying marine fauna. One such was Joseph Hooker, who sailed with Sir James Clark Ross on HMS *Erebus* from 1839 to 1843. During the expedition, Hooker dredged in Antarctic waters and made pelagic collections, although his primary interest remained botany. He met Huxley before he sailed and the two became close friends. Hooker knew many promoters of ocean science personally and kept abreast of developments in the field. A friendship between his and George Wallich's fathers prompted Wallich to keep Hooker apprised of his *Bulldog* work. Hooker also knew Carpenter. The year of the first *Porcupine* voyage, Carpenter invited Hooker to attend a presentation on the *Porcupine* results, noting "how highly you esteem both Eozoa and our Deep Sea researches." Hooker contributed directly to the organization of the *Challenger* expedition in several ways. He not only sat on the Royal Society committee and recommended Henry Moseley for the position of assistant naturalist, but he refereed by mail the disagreements that arose between Moseley and Thomson over the amount of botanical collecting that the ocean-oriented expedition should undertake. He became president of the Royal Society in 1873, during the expedition, and served as the official conduit through which Carpenter and Jeffreys approached the Admiralty for several post-*Challenger* ocean science cruises.[75]

The well-known physicist William Thomson likewise lent his support to the Royal Society's efforts. His interest in the ocean derived from involvement in the Atlantic cable enterprise. He was acquainted with the work of naturalist-dredgers from his visit to Carpenter at Holy Island. In 1870 he purchased a yacht, the *Lalla Rookh,* which became an integral part of his social, business, and intellectual life. His experience as a veteran of several cable-laying voyages and his interest in making navigation safer prompted him to tackle the problem of accurate sounding equipment. In 1872, he designed a wire sounding device that he tested at sea from his yacht. Unlike earlier experimenters, who used iron or copper wire, Thomson used steel pianoforte wire, which became readily available in the second quarter of the nineteenth century, when pianos became affordable to the middle classes. He also brought together into a single framework all the rigging and equipment for paying out, measuring, and reeling in the wire. Instead of being spread across the deck, his sounding device was a self-contained machine. He developed his sounder commercially, selling one version of the machine, whose wire was 300 fathoms long, for navigational sounding and another, bigger machine for deep-sea sounding.[76]

The support of articulate, visible spokesmen for ocean science added power to the already compelling set of arguments for deep-sea investigation. The Admiralty, in complete accord with the Royal Society's proposal, responded swiftly and favorably in April 1872. The entire expedition was organized in only nine months, a testament to the long preparation by the Royal Society committee and the cooperation of the Hydrographic Department of the Admiralty. The ship selected for the voyage, HMS *Challenger,* was large enough to carry a scientific staff, heavy machinery for sounding and dredging, other equipment, and an enormous amount of rope. Renovations to the vessel included the removal of all but two of its guns to make room for laboratories and extra cabins for scientific staff. Nares, who had sailed with Carpenter on *Shearwater* the previous summer, was chosen as commander.

As with the *Lightning* and *Porcupine* voyages, the Royal Society appointed the civilian scientific staff. At the age of fifty-nine, Carpenter was too old to accompany the expedition, leaving Thomson the obvious choice for chief scientist. Three other naturalists were chosen: Henry Nottidge Moseley, who had already abandoned medical studies to participate in the 1871 eclipse expedition to Ceylon; John Murray, who likewise studied medicine at Edinburgh University but had left to join a seven-month Arctic whaling voyage; and Rudolph von Willemöses-Suhm, a young German biologist who visited Edinburgh with the German North Sea Expedition in 1871 and there met Thomson. A chemist, John Young Buchanan, and an artist, John James Wild, rounded out the party.[77]

Scientists, officers, and crew members joined *Challenger* at Shearness, where the ship was fitted out and equipped for its special scientific work. The Lords of the Admiralty and members of the Royal Society committee inspected the ship and shared a farewell dinner with scientists and officers, then the *Challenger* left for Portsmouth. From there the expedition departed on December 21, 1872. Three and a half years, 68,890 nautical miles, and 362 scientific stations later the *Challenger* returned. During the cruise, Huxley's mysterious *Bathybius* was debunked as a precipitate of calcium sulfate, formed by the reaction of preserving alcohol with seawater. Aside from this disappointment, the expedition was a stunning success. It brought back about 7,000 specimens, half new to science, and set records for sounding, down to 4,475 fathoms, and dredging, from over 3,000 fathoms. To all observers the expedition reflected national glory and the excellence of British science.[78]

ARGUMENTS ABOUT the scientific promise of the sea, however strong, were not as persuasive to the British government as the tangible prospect that ocean science would promote submarine telegraphy. Equally convincing, and closely connected, were appeals to national pride.[79] Although deep-sea work by hydrographers and scientists in the United States had ceased during the Civil War, they resumed their efforts after hostilities ended. To the eyes of British ocean scientists, this work threatened not only British dominance of this emerging field but also Britain's identity as a maritime nation.

Professional scientists interested in the ocean urged government support for their research. The science of geology, which enjoyed significant state patronage, seemed a relevant model, especially because of the tight intellectual connections between dredging and geology. When British ocean scientists began to lobby for government funding, they promised results of great scientific value that would carry "us back into the remote past." Geographic exploring expeditions, such as the famous Arctic and Antarctic expeditions of the first half of the nineteenth century, also seemed to argue for similar assaults on the unknown depths of the sea. In 1863, John Gwyn Jeffreys characterized BAAS grants as an effort to compensate for the shortcomings of the government. He made explicit the usually implicit connection that ocean scientists began to draw between dredging as a means to study the sea and the "exploring expeditions such as used to be undertaken." When the Royal Society approached the Admiralty with its circumnavigation proposal, it had in mind the recent example of government-funded eclipse expeditions in December 1870 and 1871.[80]

By invoking expeditions of geographic exploration and discovery, ocean scientists hoped to reinforce the idea that the sea ought to remain the purview of the British Empire. This analogy suggests why British naturalists advocated a major circumnavigation rather than more small-scale voyages like those of the *Porcupine.* To encourage government funding, ocean scientists presented their work to the public in nationalist terms, as when Jeffreys wrote in *Nature,* "I feel confident that Great Britain, with her vast wealth, naval resources, intelligence, energy, and perseverance, will keep the lead which she has now taken."[81]

At the time that Carpenter, Thomson, and others broached the idea of the *Challenger* expedition, they also tried to shame the government into funding the expedition by pointing out what the Americans were doing. As the *Lightning* and the *Porcupine* sailed, the U.S. Coast Survey was conducting deep-sea sampling in the Gulf Stream. The Survey had long patronized ocean science through resources it extended to Louis Agassiz, and later his son Alexander. In 1868 Louis Agassiz and Louis F. de Pourtales began the Gulf Steam study using the Coast Survey steamer *Bibb* and later the *Hassler.* British scientists referred to dredging work from *Bibb* as "the

American Deep Sea Exploring Expedition," hoping that the unfavorable national comparison would generate support for British ocean exploration.[82] But because the Coast Survey had been undertaken for economic reasons, neither Coast Survey officials nor Congress would have agreed with or appreciated British scientists' characterization of this work as the equivalent of a major national scientific expedition.

British scientists argued that their government did not support any branch of science as extensively as did the United States and countries on the Continent. For ocean science in particular, they warned of American preeminence. William Carpenter lamented dramatically that his country's status as the premier maritime nation was endangered because its scientists lagged behind Americans in conducting "Marine Surveys." In the pages of *Nature* he invoked strongly nationalistic rhetoric, warning, "We are falling from the van into the rear . . . Is this credible to the Power which claims to be mistress of the seas?"[83]

THE CHALLENGER expedition has been remembered as a landmark national accomplishment and the genesis of the science of oceanography. Although its fifty-volume report arguably serves as the intellectual foundation of the field, the expedition itself represented the culmination of midcentury questions, practices, and traditions of ocean investigation. It was novel only in its huge scale. The *Lightning* and the *Porcupine* had already knit together the work of deep-sea hydrographers with that of naturalist-dredgers. The instruments carried aboard *Challenger* were mostly those tried and tested earlier, not the technology upon which future ocean science would be based. Nets, dredges, and sounding devices were deployed with hemp line, not the wire line and wire rope that were already the object of experimentation for deep-sea sampling. *Challenger* did sail with one of William Thomson's sounding machines but, after the drum collapsed the first time hydrographers used it, they abandoned it and relied on proven gear.[84]

Although the *Challenger* sailed with conservative technology, the sailors, officers, and scientists continued the tradition of shipboard tinkering to improve collections. John Murray, who became a major proponent of the new science of oceanography, experimented frequently with equipment, trying to use nets to collect in intermediate depths, employing sounders to display in cross-section layers of ocean floor, and experimenting with dredges used by native fishermen along coasts the expedition visited. Other crew members began to employ Calver's hempen tangles alone,

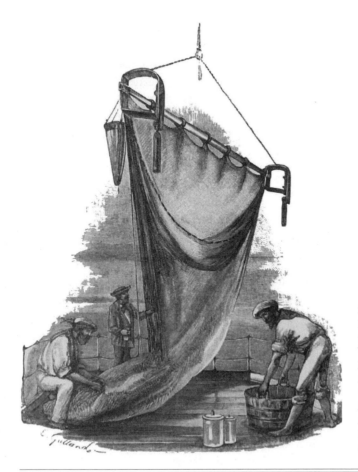

Sailors emptying the trawl on HMS *Challenger* while a scientist looks on. (From Charles Wyville Thomson and John Murray, ed., *Report of the Scientific Results of the Exploring Voyage of the HMS "Challenger," 1873–76: Narrative of the Cruise*, vol. 1 [London: HMSO, 1885], 236; courtesy of Dr. David C. Bossard.)

rather than attached behind a dredge, when they realized that, on rocky surfaces where dredges could easily get stuck, bunches of swabs skimmed over the ground, entangling echinoderms and other creatures.[85]

Captain Nares was responsible for another innovation that drew on material resources available on board. One day during the third month of the expedition, he faced a dredge haul that came up full of ooze containing nothing of scientific interest. Frustrated, he ordered that the dredge be replaced for the next haul by a compact fishing trawl thirty feet long with a thirteen-foot beam. This first attempt at deep-sea trawling brought up a rare specimen, which attracted the scientists' attention. Experiments continued and the trawl proved successful in 600 fathoms, then 1,090 fathoms. Eventually trawls were employed at all depths, the deepest successful *Challenger* trawl being 2,650 fathoms. Trawls skimmed the surface of the sea floor, collecting only loose objects on top rather than digging down into mud as the dredges did. They could be dragged along the bottom for up to a mile, returning full of animals. The expedition artist was especially delighted by the trawl, which recovered more "perfect" specimens than swabs. Officers and men saw little distinction.

> There's a Trawl for fancy drudging, and the work's about the same,
> The only difference I can see, is that wot's in the name.

If anything, trawling was more tedious for the crew than dredging, because dragging the trawl over so much ground translated into extra hours spent keeping the net out. In very deep water, trawling from the *Challenger* took twelve to fourteen hours.[86]

While *Challenger* made record-breaking collections with traditional equipment, American ocean investigators pioneered the next generation of dredging technology. They did so not in the context of a major expedition, but aboard working hydrographic vessels. Although the *Challenger* staff eschewed Thomson's wire sounding machine, it had attracted the attention of the chief of the U.S. Bureau of Navigation, and two American naval officers adopted it immediately. Captain George Belknap took one of Thomson's machines aboard the USS *Tuscarora* for a cable survey between California and Japan in 1873 and 1874. During a shakedown cruise Belknap "determined the superiority of Sir William Thomson's machine and piano forte wire over the steam reel and rope." He did not adopt Thomson's machine as it was, but made "such alterations and improvements . . . as were suggested by experience." His assessment was enthusiastic: "the moment of touching bottom at 4600 fms was as distinctly, instantly, and accurately known as at 1000 fms or 100 fms. That may seem strange, but it is a fact and I never tire of the wonderful working of the machine."[87]

During *Tuscarora*'s work in the Pacific, Belknap had the opportunity to spread the word about wire sounding to another American ocean investigator. William Healey Dall was a naturalist of significant technical ability who became so involved in his surveying duties that his mentor Spencer Baird teased him about losing interest in natural history. He defended himself: "Do not imagine for a moment that I have any idea of branching off into surveying work except . . . for this occasion only. The moment I have obtained the material I want, I shall . . . come back to Washington at once." Yet Dall welcomed the chance to see Belknap's machine and reported with excitement to his boss that Belknap had devised the best sounding gear he

The sounding machine devised by Captain Charles D. Sigsbee on the U.S. Coast Survey steamer *Blake*. Sigsbee and Alexander Agassiz worked together to apply wire rope to deep-sea dredging. Sigsbee then wrote what became the standard reference work on deep-sea equipment. (From Charles D. Sigsbee, *Deep Sea Sounding and Dredging* [Washington, D.C.: Government Printing Office, 1880], fig. 3; courtesy of the National Oceanic and Atmospheric Administration, U.S. Department of Commerce.)

had ever seen. He resolved to "[get] up a sort of compromise sounding apparatus more fitted for a sailing vessel," then secured permission to use his new vessel, the *Yukon,* to test his version of the gear.[88]

Although the Coast Survey used Dall's drawings to build a copy of his device, it was not widely adopted, probably because the use of sailing vessels for survey work was rare and diminishing. Belknap's adaptations did survive; Thomson himself adopted his substitution of a set of weights instead of tackle to put tension on the brake rope. Another American, Captain Charles D. Sigsbee, designed another adaptation of Thomson's sounding machine, while commanding the Coast Survey steamer *Blake.* Sigsbee was more well known as commander of the *Maine,* which mysteriously exploded in Havana Harbor in 1898 while Cuba was struggling for independence from Spain. In 1880 after a series of cruises on the *Blake* with Alexander Agassiz, Sigsbee wrote a book that became the standard reference on deep-sea sounding and dredging technology. As a result of contributions by Thomson, Belknap, and Sigsbee, cable companies began to build self-contained machines capable of making flying soundings (measuring depth while continuing to steam ahead) in less than 200 fathoms. In deeper water, ships hove to while lowering the sounding line, but steamed ahead while reeling it in. This technology reduced sounding time to 40 minutes, from start to finish, for 2,000 fathoms. In 1887, the British telegraph engineer Francis Lucas patented a new sounder that became standard equipment on cable industry and Royal Navy vessels.[89]

Aboard the *Blake* Sigsbee, working with Agassiz, pioneered the use of wire rope for deep-sea dredging. Agassiz praised Sigsbee: "To his inventive genius is due the efficient equipment of the Blake and his suggestions have modified all the apparatus originally in use on the vessel." Agassiz's technical ability was an essential element of the partnership, though. He enjoyed a close association with the Coast Survey, an inheritance of sorts from his father, and worked during 1859 as an aid to a Coast Survey party on the California coast. His 1855 civil engineering degree from Lawrence Scientific School proved helpful, as did his experience of managing the Calumet and Hecla copper mines in northern Michigan, where he made the fortune he would later spend liberally on oceanographic research. Perhaps his mining experience suggested to him the replacement of hemp with wire rope for dredging. Invited by Coast Survey Superintendent Carlile P. Patterson to serve as scientific director of deep-sea dredging operations on a series of Coast Survey cruises, Agassiz apprised him of his ideas. The two men discussed with Sigsbee and his officers ideas for equipment that was designed, built, and then tested on the first *Blake* voyage in 1877.[90]

American hydrographers and scientists took pride in technological improvements, just as they did in the data and specimens their equipment enabled them to acquire. National competition characterized deep-sea research as it started to become an established branch of science. While the *Challenger* expedition sailed, Dall wrote from Alaska to his Survey boss, Superintendent Benjamin Peirce, that he had found globigerina mud—recent chalk formations—accumulating at a depth of 800 fathoms. This fact, he crowed, "our English fellow laborers in the Challenger would doubtless consider as a feather in their caps if they had been the ones to establish it." Belknap also compared his results to those of the famous British expedition, boasting to Dall about the results of a Pacific cruise in *Tuscarora*: "Here are 10 consecutive casts which so far as known puts the 'Challenger' quite in the shade." So sensitive were feelings about accruing national credit that, although Jeffreys and his friend and fellow conchologist Alfred Merle Norman were invited on the 1881 French *Travailleur* expedition to give advice, they were not permitted to work on the collections. Jeffreys dismissed the rebuff lightly, secure in the strength of British marine zoology: "I thought you [Norman] would be amused by the cool way in which the French Savants have appropriated all the knowledge they gleaned during the cruise of the 'Travailleur.' Never mind. We can afford it."[91] The *Travailleur* voyage was one of many undertaken by maritime nations following *Challenger*'s example. Others include the voyages of the French *Talisman*, the Dutch *Siboga*, the Danish *Ingolf*, and the German *Valdivia*.

Yet despite the attempts of individuals to outdo each other, the final quarter of the nineteenth century saw the rise of international sentiment and action, in which economic, technical, and scientific sectors were organized across national boundaries. When the time came to assign the vast *Challenger* collections to the scientists who would study them, an enormous controversy erupted between people who advocated choosing only British scientists and those who argued for selecting the "best" scientists in each particular field, regardless of nationality. When the *Challenger* Office was established in Edinburgh to oversee the production of fifty volumes of reports, its director, the newly knighted Sir Charles Wyville Thomson, adopted the international approach. John Murray took over the office after Thomson's death, until the last report appeared in 1895. The *Challenger Reports* summarized all natural, physical, and chemical ocean science to that time. Thereafter, the ocean—including its greatest depths—remained the province of scientists from a wide spectrum: zoology, geology, chemistry, and physics.[92]

chapter six

Small World

"That science from our deeds may grow."

—From "A Song for the 'Challenger' Crew," printed in Philip F. Rehbock, ed., *At Sea with the Scientifics: The "Challenger" Letters of Joseph Matkin,* 1992

CHALLENGER neither set a precedent for government patronage of ocean science nor single-handedly established this field as the successor of grand national exploring expeditions, but its voyage did leave a lasting impression. Producing scientific knowledge about the deep ocean during the nineteenth century required scientists to integrate their work with existing maritime practices, technologies, and social settings that were suffused by a masculine naval culture and subject to the physical challenges of constrained space, harsh discipline, bad weather, and worse food. Scientists traveling on ships whose primary mission was not science fit uneasily into the ship's bounded universe. Relationships between scientists and sailors were characterized neither entirely by antagonistic conflict nor solely by peaceable cooperation. Instead, a new culture of scientific work at sea was forged from scientists' entry into the established and functional social order of working vessels, from the exploring expeditions of the early nineteenth century through *Challenger*'s voyage. This culture represents a fusion of mid-nineteenth-century scientific and maritime work cultures.

The act of going to sea defined practitioners of early ocean science more than a shared body of specialized knowledge or common methods. Research at sea gave ocean scientists similar experiences, even when their intellectual affiliations were as diverse as geology, microscopy, zoology, and hydrography. At the same time that investigators began probing great depths, similar efforts were under way to explore other vast geographic areas, including the Arctic, the atmosphere, and mountains. Gripped with the sense that all of the earth's land and islands had been discovered and claimed, explorers turned to these last frontiers. Anthropology, too, appeared at this time to document fast-disappearing peoples and cultures. By the end of the nineteenth century, armchair scholarship, based on observations by missionaries, travelers, and colonists, had given way to the need for experts to go out into the world, so that fieldwork became a defining feature of all professional naturalists.[1] The common experience of seagoing likewise served to distinguish oceanographers from other scientists. The distinctive style of seagoing scientific work that emerged in the nineteenth century provided a culture for the field research of oceanography as its practitioners began to define it as a discipline.

When men of science encountered maritime culture, the maritime world itself was undergoing dramatic change. Long before the nineteenth century, an international maritime culture existed that polite society steadfastly ignored. On everyone's social scale, including that of the very poor, common sailors were the lowest of the "lower sort." From at least the seventeenth century on, ships' crews were international assemblages, comprised of men caught "between the devil and the deep blue sea." To such people, even life as a pirate or an outlaw might seem preferable to working under a cruel captain. From the perspective of captains, rigid discipline was the only way to contain the danger of mutiny and channel the men's energy to the work at hand. As late as the 1870s, boys accused of vagrancy and slight misdemeanors were sentenced to a year of training for jobs as able-bodied seamen on the school-ship *Mercury*, run by the Commissioner of Hospitals and Prisons of New York City. Well into the nineteenth century, the sea was widely held to be an appropriate calling for the poverty-stricken, orphaned, criminal, and insane.[2]

Life at sea changed significantly during the nineteenth century, not only for the fashionable travelers who flocked to book passage on speedy and luxurious steam liners but for emigrants and mariners as well. Reform efforts in the first half of the century made the sea a relatively more benign workplace. In both the United States and Britain, reformers publicized the conditions of life and work for mercantile and naval crews. Fearing for the souls of sinful sailors, churches set up missionary-type operations near docks. Many of the writers who went to sea, most notably Richard Henry Dana, intended their novels to educate the public about the deplorable living and working conditions of common sailors.[3]

Although growing middle-class awareness of the conditions of life at sea galvanized reform efforts, people of all social classes maintained traditional assumptions that seamen were degenerates and that life at sea was uncomfortable and undesirable for polite, repectable people. During the *Rattlesnake* voyage, Thomas Huxley discussed the evils of ship life with Second Lieutenant Dayman as they walked the decks during watch. They agreed that it was "the worst & most unnatural . . . fit for none but the unscrupulous . . . it [of all courses of life] tended most to harden the heart & render the conscience callow." George Wallich described a sailor who had laughed off a severe flogging as an "incorrigible character." As late as 1873, Lieutenant Herbert Swire wrote in his *Challenger* journal that he could not believe the general populace admired and were amused by common sailors. Hardly "jolly tars," Swire insisted, sailors were, on the contrary, "very disagreeable, and only show jollity in the occasional perpetration of objectionable practical jokes." In spite of the extension on board ships of the practices and ideals of urban planning and middle-class reformers, the sea effectively resisted encroachment of land law well into the second half of the nineteenth century.[4] To many observers, the sea remained an appalling place, the refuge of degenerates and a dangerous, immoral environment.

Although disdain of common sailors did not disappear entirely, the mid-nineteenth-century discovery of the sea as a romantic and heroic place diluted disgust with sea life and added a measure of admiration for the open ocean. The vogue of the seashore attracted middle-class attention to the ocean and members of polite society began to express cautious interest in ocean travel. Tentatively, they sampled life, and even work, at sea. Yachts, packet ships, and steamers bore first aristocrats and gentry, and subsequently the middle classes, out to the blue waters. Ralph Waldo Emerson expressed well their trepidation, which gave way to enthusiasm: "I find the sea-life an acquired taste, like that for tomatoes and olives. The confinement, cold, motion, noise, and odour are not to be dispensed with."[5] Writers imbued the act of going to sea with new meaning, creating the expectations that generations of passengers and sailors took with them to sea. As Emily Dickinson put it,

> Exultation is the going
> Of an inland soul to sea,
> Past the houses—past the headlands
> Into deep Eternity
>
> Bred as we, among the mountains,
> Can the sailor understand

> The divine intoxication
> Of the first league out from land?[6]

Dickinson's words reflect the extent to which, by midcentury, seagoing was a new experience, one freighted with deep meaning, especially for first-time voyagers.

Fashion alone did not attract people to the uncertainty of riding in cramped quarters over boisterous seas. Had ocean travel not become demonstrably safer during the first half of the nineteenth century, trend-setters would have remained content to watch the surf from the beach. With the advent of reliable steam-powered vessels in the 1840s and 1850s, as well as navigational advances and healthier provisioning for long voyages, sea travel became predictable in duration and fairly comfortable. Although the possibility of shipwreck loomed large in the minds of passengers, and actual losses were publicized lavishly, in fact sea travel had become a relatively safe undertaking by midcentury.[7]

As the sea became safer and sailors marginally more respectable, the act of setting sail on the blue water shifted from unfortunate necessity into a heroic undertaking. The sea, in short, became a personally meaningful destination. Middle-class men of science, who embraced the midcentury values of bravery and manly sport, followed naturalist-explorers, yachtsmen, and professional writers out to sea. Once aboard oceangoing ships, they participated actively in the process of redefining seagoing as an arena for challenging nature and achieving personal growth. Simultaneously, they helped renovate maritime culture to make it more familiar and comfortable for themselves and more effective in promoting their scientific work.

Gentlemen of science who crossed gangplanks to board naval vessels entered a world that was physically, spatially, and socially alien. Decks felt different underfoot than terra firma. Vessels were absolutely bounded by water, which reinforced landsmen's sense of isolation and vulnerability. The fortunate scientists who became accustomed to life on ships could work as easily on board as off. The *Challenger* scientist Henry Moseley reflected, "It is wonderful how completely practice enables a man so to modify his movements so as to perform with success, in a ship constantly in motion, even the most delicate operations. The adjustments of the body to the motion of the ship in ordinary weather become, after a time, so much a matter of habit." Of course not everyone adapted to shipboard life with such aplomb, nor was the weather at sea always ordinary. A few miserable scientists never got their sea legs. Once they regained shore, though, they converted their uncomfortable experiences into tales of heroic persever-

to coal. say four days - and then steam straight for Portsmouth. In all probability, therefore we shall reach that port by the 28th or 30th inst.

The skipper has told me this morning that in all his experience he never encountered such a terrific storm as that of Monday & Tuesday. That even in the heaviest one he has been in, they have never extended over nearly so long a time as this last one did, namely thirty hours. I mentioned to him the remarkable coincidence of the date being the same as that on which the Royal [Theatre] was lost last year namely the 8th Oct., at which he seemed much struck. The skipper took me to look at the broken stump of one of the davits of the cutter which was lost. It may afford some idea of the terrific force of the sea which swept the boat away, when I state that the solid rod of iron - five inches thick, has been snapt across close to the bolts as if it had been a thread.

Whilst I write I am clinging on by my left hand to my table and supporting my body by stretching my legs out against my chest of drawers, in order to enable me to steady my right hand at all. During the heavy lurches it is all one can do to hold on at all without writing. On deck the ship appears as if she were about to roll entirely over.

So much for a sketch done under the circumstances. All I can say is that it does not give half the idea of the angle at which the ship frequently rolls over - or of the tremendous volume of sea

A sketch by George Wallich on a page of his journal from the voyage of HMS *Bulldog,* showing the sidewheel paddle steamer facing huge waves in a storm. (From Wallich, "Daily Diary of the Voyage of 'Bulldog,' No. 2," Oct. 4, 1860, in the George C. Wallich Papers, Natural History Museum, London; © the Natural History Museum, London.)

ance. Alexander Agassiz's grandson described how his grandfather overcame seasickness: "Anyone afflicted with the malady can easily imagine what fortitude and enthusiasm it must have required to crawl on deck from a bunk of despondency and pain and lose one's self in the eager examination of the treasures which the dredge had just brought to the surface."[8]

Even for the stalwart, storms were a pervasive theme in their reflections. George Wallich drew a sketch of the *Bulldog* in storm waves and wrote in his journal, "The sea was lashed into such a vast cauldron that it presented but one unbroken sheet of white foam hiding the horizon entirely from view . . . All I can say is that [the sketch] does not give half the idea of the gulphs of seas she [the ship] has constantly to encounter." Scientists marveled at sailors' ability to work aloft under the worst conditions. "To a landsman . . . it appears like courting certain destruction when a man 'lies out' as it is termed on the extreme end of the yard arm . . . Yet the sailor goes about the task as readily & unconcernedly as if the risk were nothing." Landlubbers often focused on details that would not ordinarily have appeared in more salty reminiscences, revealing their tendency to stay below. On HMS *Rattlesnake* during a gale, Thomas Henry Huxley wrote, "Every now and then . . . there is an instant of silence, then comes a roll. Ugh, the timbers creak, the pigs squeal, the fowls cackle, two or three plates fly with a crash out of the steward pantry."[9]

At the same time that ocean scientists faced the novel terror of ocean storms, story tellers, especially Richard Henry Dana, began to depict as heroic the sailors who braved tempestuous Cape Horn. Scientists at sea interpreted their experience of rough weather in light of such accounts. When they published their own voyage narratives, they contributed to the valorization of seagoing men, specifically those who braved the ocean's dangers for science.

Even when sailors were spared horrific storms, the sea was often too rough for scientific work. Scientists had to learn that their collecting activities proceeded at the mercy of Neptune. Safely deploying apparatus over the side required a relatively calm sea since excessive rolling and pitching caused lines to snap, resulting in equipment loss. In his field notebook, the U.S. Coast Survey officer and marine zoologist William Healey Dall noted, on choppy days, "Wind too high for field work." In the midst of an almost decade-long study of Shetland fauna, Alfred Merle Norman despaired, "Oh, for fine weather! If we could but get it we might yet do well and get good things to reward us . . . It is very tantalizing. Here we are with every possible appliance, ready . . . yet the weather is such that we might as well be in England." In this case, as in most, Norman lamented stormy weather, but very calm days could be equally deleterious for scientific work. On one occasion, the Shetland dredgers waited on their yacht for hours, with "not a breath of wind." Finally, they had the vessel towed out to sea, only to encounter a dense fog that made it difficult to find their dredging site. In spite of the calm, there was a strong tide and heavy swell on the sea, so that, as John Gwyn Jeffreys reported to Norman, "Our grand expedition, so anxiously prospected, was almost a total failure."[10]

Aside from the extreme discomfort of storms, the most unsettling physical aspect of daily life at sea to landlubber scientists was the rigid and hierarchical arrangement of space on ships. Close quarters encouraged a theatrical quality to life aboard ships, where virtually all places, actions, and speech underwent public scrutiny and interpretation, so that space, rank, and behavior were related. After its port stop in Tahiti, HMS *Bounty*'s main social space was appropriated to transport breadfruit trees. Subordination of this space to the ship's botanical function contributed significantly to the famous mutiny. Similarly, problems arose for ships in which science was appended to, or integrated into, a more traditional mission. Prime cabin and deck space was given over to scientific work. Laboratories and workrooms were built on deck, so that scientists had as much light as possible for their work. This contrasted starkly with the work locations of sailors such as Joseph Matkin, who kept the steward's books in a small, dark room several levels below *Challenger*'s deck.[11]

On HMS *Porcupine*'s 1869 expedition, Captain Calver had to rearrange space so that the visiting scientist and two assistants would be physically accommodated according to their social status relative to the naval hierarchy. His plans included housing two scientific assistants in a lower-deck cabin previously occupied by midshipmen (9 feet × 5.5 feet for two men). Extra space in that cabin ("extra" defined by naval standards) could be used for specimen storage. The assistants would mess with either the engineers or the midshipmen, while the chief scientist would mess with the captain. Calver suggested two possibilities for housing the chief scientist. One involved removing the chartroom table entirely and putting in that cabin "a handy small sized table for the examination and preparation of specimens, and . . . a couple of cots and working stands" so that the space would accommodate both the scientist and an assistant surgeon. Alternatively, he offered to appropriate part of his own cabin for "the FRS afloat." It remains unclear exactly where everyone finally slept on the *Porcupine,* but the chartroom was indeed transformed into the laboratory.[12]

Whereas on the *Bounty,* space for its scientific mission came from the crew's social area, *Challenger*'s scientific work appropriated space from areas formerly used for military purposes. Sixteen of the eighteen sixty-eight-pound guns were removed. The main deck was reserved for the equipment and operations of deep-sea sounding and dredging, but a platform erected above the center of the upper deck allowed sailors and scientists to dump, sort, and sieve the contents of the dredge without hampering deck work. Open space on the main deck was devoted to hoisting machinery and the storage of gear too large for the workrooms.[13]

Within the vessel, space was allocated on the principle that the scientific work was parallel to the more traditional navigation and surveying work. The radical step of dislodging the captain from his traditional solitary place in the aft cabin revealed the central importance of science. Divided in two, the aft cabin housed Captain George S. Nares and Professor Charles Wyville Thomson in parallel positions of power. They also shared the 30 foot × 12 foot fore-cabin, which had skylights and a writing table, as their personal sitting room and study. Reification of science in the ship's structure emphasized to mariners that *Challenger* was like no other vessel they had sailed.

While Nares and Thomson shared the fore-cabin as work space, naval and civilian scientific staff had congruent but separate workplaces. Naval officers made hydrographic, magnetic, and meteorological observations and oversaw the operations of sounding, dredging, and temperature measurement from their headquarters in the starboard chartroom. The natural-history workroom stood opposite that, on the port side. There, scientists engaged in the same work as they would have done in museums or physiology laboratories, modified specifically for marine organisms. The naturalists' workroom was only two-thirds the size of the fore-cabin shared by Nares and Thomson, although it was comparable in size to the chartroom.[14] Size discrepancies such as that between their workspace and Thomson's starkly introduced the junior naturalists to the meaning of rank translated into spatial terms.

Just as public and work space were clear indicators of the rank and importance of the activities undertaken there, so too was personal living space, which was always at a premium on ships. Crew members had almost none, so one of the most coveted privileges of rank was having a cabin, even a shared one. On most naval vessels, junior naturalists shared bunk space in the wardroom, which put them on a par with midshipmen, or junior officers. Naturalists lived in different places on ships depending on their official positions. Surgeon-naturalists, who were ranking naval officers, had cabins that doubled as examination rooms, usually located near the area designated for sick sailors, if one existed. Men who, like

The HMS *Challenger*, showing the platform erected to create a place for scientists to work without hampering the work of the ship's crew on deck. An officer supervised the machinery for dredging and sounding, relegating scientists to the role of observers until the nets were brought aboard. (From Charles Wyville Thomson and John Murray, ed., *Report of the Scientific Results of the Exploring Voyage of the HMS "Challenger," 1873–76: Narrative of the Cruise*, vol. 1 [London: HMSO, 1885], 57; Courtesy of Dr. David C. Bossard.)

Charles Darwin, were invited on voyages to keep the captain company, usually slept in the captain's cabin, which was also where wealthy passengers stayed before ocean travel became regularized enough to convert cargo space to passenger cabins. Government expeditions that carried official naturalists provided them with cabins that also served as workrooms.

Multipurposeness was a pervasive feature of naturalists's accommodation. Although officially Charles Darwin shared Captain Robert Fitzroy's cabin, in practice he apparently used other quarters. One of Darwin's *Beagle* shipmates wrote to Joseph Hooker to lament that none of them had been able to attend Darwin's funeral to represent his time at sea. He reminisced: "The narrow space at the end of the chart table was his only accommodation for working, dressing, and sleeping, the hammock being left hanging over his head by day when the sea was at all rough, that he might lay in it with a book in his hand when he could not any longer sit at the table. His only stowage for clothes being several small drawers in the corner reaching from deck to deck; the top one being taken out when the hammock was hung up: without which there was not length for it."[15] This description shows the preciousness of space aboard ships and the extent to which Darwin had to accommodate his study and sleep to the primary navigational and hydrographical work of the *Beagle.*

William Stimpson, the North Pacific Exploring Expedition naturalist, had a small cabin and a voracious appetite for collecting, so that he frequently had to overhaul his room and send boxes of specimens to the hold or to the expedition ship designated to carry supplies. The small size of his cabin was exacerbated by the fact that he shared it with a lot of collecting gear, including four scoop nets and three wire sieves as well as his microscope, dissecting instruments, and black saucers used to examine transparent animals. By contrast, navigation and surveying equipment would have been assigned space on deck or in deck cabins. Stimpson also kept all of his own drawing and writing equipment in addition to numerous kinds and sizes of jars and vials for storing and sorting specimens. In spite of tight quarters, though, when the temperature plummeted, he and his botanical colleague Charles Wright found "it comfortable to shut ourselves up in our rooms, where writing and study are prosecuted with that zest which cold weather always gives to such pursuits." But when the wind and sea picked up, "The rolling and pitching of the ship [gave] us no chance of writing or working, and scarcely any of reading." Officers and crew, of course, did not have the option of staying below in icy weather or heavy seas.[16]

George Wallich, who had lobbied energetically for his position on HMS *Bulldog,* expressed frequent disappointment with his living and working

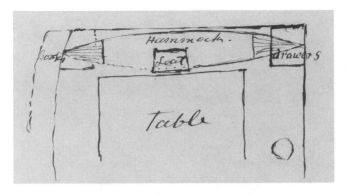

A drawing by a shipmate of Charles Darwin's of Darwin's work and sleeping space on HMS *Beagle*. (From B. J. Sulivan to Joseph Hooker [c. 1882], Charles Darwin Papers, Cambridge University Library, MS.Dar.107:f.45r; reproduced by permission of the Syndics of the Cambridge University Library.)

conditions during the four-and-a-half month voyage. Initially, he complained that his equipment was scattered all over the vessel, with the exception of his dredges, which he put in his cabin to have immediate access to them. Two weeks later, Wallich brought the chest of specimen bottles into his cabin as well, because he feared that the crew members who had to share their space with it were knocking it around. Soon, like Stimpson, he lived with virtually all his collecting and preserving equipment and specimens in his cabin. The second day of the cruise, Captain Leopold McClintock invited Wallich to use his cabin whenever the light in Wallich's quarters or the wardroom was insufficient. Although Wallich managed to do microscopic examinations in his own quarters, he complained about the "contaminated and filthy air" around his cabin. This, he believed, was caused by its location, "in common with the Stewards pantry, the Dispensary and the Assistant Surgeon's quarters," especially when the area was used either for imprisoning drunken sailors or as a sick bay.[17]

Not all ships' naturalists were unhappy with their accommodations, and many became attached to their small domains. Henry Moseley did not suffer in the least from confinement on the *Challenger*. In fact, he welcomed the freedom from the day-to-day distractions of newspapers and letters so he could devote his time to work and reading. He memorialized his shipboard home in a sketch and reflected, "I felt almost sorry to leave, at Spithead, my small cabin, which measured only 6' × 6', and return to the more complicated relations of 'shore-going' life, as the sailors term it. I had lived in the cabin three years and a half and had got to look upon it as home."[18]

Next to personal living space, the biggest preoccupation of both scientists and sailors was food. For common seamen, food as well as their grog

Henry Moseley's sketch of his cabin on HMS *Challenger*. Mosely also made a watercolor from this sketch, which is in Moseley's albums in the Department of Zoology, University of Oxford. (Edinburgh University Library, Special Collections Department, HMS *Challenger* Collection, MS Gen. 20.3; reproduced with the kind permission of the Edinburgh University Library.)

rations provided welcome respite from work, and meals served important social functions in shipboard life. Sailors looked forward to Sundays and holidays, when they were served special dishes such as plum duff, a pudding with raisins. Food thus helped mark time. Early in a voyage, the act of eating was a major difficulty, but as greenhorns got used to the sea they viewed eating as an accomplishment. *Challenger* lieutenant Herbert Swire described Christmas dinner less than a week after the vessel cleared Spithead: "We had to hang on to our grub pretty stiffly, for everything had a tendency to make for the lee side of the place, there to revel in sublime smashery and confusion."[19] Once a ship had been away from land for more than a few weeks, fresh food was gone. Large ships on long voyages carried livestock to slaughter on route, but fresh meat, when available, was reserved for the officers' mess, and even that ran out quickly. Common sailors ate salted meat. A similar split held for carbohydrates: officers often got bread while men ate ships' biscuit or hardtack.

The quality of preserved food, including salted meat or biscuits, improved during the nineteenth century as food-preservation techniques improved. Nevertheless, it remained common for long-distance sailors to find insects in their food. Instances of scurvy decreased as awareness grew about what kinds of food inhibited it. Still, on large naval vessels that carried food for hundreds of men for weeks or months, meals at their best were not usually appealing.[20]

While sailors almost ritualistically complained about food, it was gentlemen seeing ships' fare for the first time who recoiled in horror at what they saw. On the *Bulldog*, Wallich was no happier with the food than with his cabin. Early in the voyage he decided that he would think himself "most fortunate if I get back without Scurvy." Dinner that day had included "a piece of Salt Junk so tough and hard and salt as to be with great difficulty masticated." Later, he complained of being served "bread unbaked almost." Although from the educated middle class, Wallich was not without culinary experience with which to judge ships' food. He compared the *Bulldog*'s fare unfavorably to the camp food he ate during field campaigns and wars while serving in HM Indian Army. *Challenger* scientists likewise deplored the lack of fresh food, as when Thomson dolefully lamented the day the crew killed the last sheep and the scientists returned to the hated preserved meat. He noted that the captain provided some "good tinned soups" for the mess, but admitted that he himself "generally [made] up with boiled rice with sugar and sherry," in addition to "two or three glasses of port wine during the day and a little rum and water at dinner." Not surprisingly, he shared the "great anxiety" of all hands to make landfall and be able to eat fresh food.[21]

Even for scientists who acquired a taste for the sea, ocean travel remained spatially and gastronomically uncomfortable. Scientists and travelers, indeed virtually all sailors, looked forward to port stops, and eagerly awaited the ships' final destinations. Nevertheless, scientists' work space as well as their living quarters were situated in the most stable parts of the vessel, aft and amidships, where captains and officers lived on naval ships and first-class passengers stayed on packets and liners. This location constantly reminded everyone on board of the primacy of the scientific mission, which to a certain extent helped ease scientists' entry into a new world.

IF SHIPS WERE unfamiliar physical environments for naturalists, their social and cultural universe was even more alien. Tensions arose readily between scientists and mariners, but the juxtaposition was not the unidimensional one of scientists meeting sailors. Rather, scientists interacted with several overlapping maritime populations. Within the crew, a clear pecking order existed that involved a combination of such factors as age, experience at sea, rank, personality, and job assignment. Most common sailors regarded with some contempt the "idlers," as they called anyone, from cook to surgeon to visiting scientist, who did not stand watch. The captain represented a separate category from the crew, and even the officers, because of his sole responsibility for safety and the successful execution of the vessel's mission. Scientists, who were the only ones not wholly under the captain's control, often disagreed with the deck officer over when they should be allowed to collect and how much help they should expect from the crew. Officers mediated between the captain's orders and the crew's work, maintaining, however, strict social segregation from the crew. Although scientists usually worked with officers who were hydrographers, who were trained in the physical sciences, even these officers did not always sympathize with naturalists' endeavors, particularly the endless collection of inedible animals. Officers' collection of physical measurements such as temperature or depth competed for equipment, labor, and time with scientists' collection of animals or bottom sediments. Scientists had to develop different kinds of relationships at each level of the ship's social strata.

Unfriendliness toward outsiders on board was quickly reinforced by antipathy toward the extra work they caused. Initially, dredging, trawling, and sounding inspired intense curiosity about what lay beneath the waves. The *Challenger*'s first few dredge hauls attracted a crowd of "every man and boy in the ship who could possibly slip away," waiting breathlessly for a glimpse of the secrets of the depths. Instead of merfolk or monsters, nets broke the surface full of sand, mud, and a soon-monotonous assemblage of animals. Moseley admitted, "Gradually, as the novelty of the thing wore off, the crowd became smaller and smaller . . . as the same tedious animals kept appearing from depths in all parts of the world."[22]

A poem written by the *Challenger*'s chief engineer, but taking the perspective of the common sailor "Jack Staylight," equated dredging work with the most hated jobs on board:

> I'll just acquaint you Topmate, with the nature of my duty,
> And show you what a lot I've learned since last we met my beauty.
> I joined this outfit last winter and got rated on her ledger
> A swabber, jobber, scrubber, a sounder and a dredger.

A vast difference existed between the standards of cleanliness and order on a naval ship and those of vessels involved in whaling, fishing, or natural-history dredging. Alexander Agassiz described his departure from the Coast Survey steamer *Blake* as "an event which must have been a relief to the officers, more particularly to the executive officer, Lt. Ackley, who was once more free to put the ship in an orderly condition." He added, with great understatement, "The work of dredging is not conducive to cleanliness." Yet despite the filth and grueling physical labor involved, some of the common sailors expressed interest in the scientific results of their dredge hauls. On the *Challenger*, Joseph Matkin appreciated Thomson's initial lecture to the crew about their scientific mission. He resented that the scientists never again repeated this courtesy during the three-and-a-half-year voyage.[23]

Naval officers, engineers, and seamen uniformly referred to the scientific work as "drudging." Especially as the voyage wore on, they grew "tired and anxious to get home." Thomson found it "difficult enough to reconcile them to extra work." Moseley admitted that "the ardour of the scientific staff even abated somewhat, and on some occasions the members were not present at the critical moment, especially when this occurred in the middle of dinner time, as it had an unfortunate propensity of doing." Only Thomson's enthusiasm never flagged; he always attended the arrival of the net at the surface and remained determined never to make the next port stop without first having caught "something good."[24]

While mariners who were forced to conduct deep-sea dredging resented the work, those watching from another vessel found it incomprehensible. In his narrative account of the 1870 *Porcupine* voyage in the Mediterranean, Captain Calver recorded the confusion caused by their cruise track. At one point, he learned from local British colonial officials that observers of their activities close to the African shore judged *Porcupine* "to be a Prussian vessel supplying the disaffected Arabs with arms and ammunition." The vessel's movements had been "communicated by Telegraph . . . and a Man of War and 250 Troops had been dispatched in search of her accordingly." While some interpreted *Porcupine*'s behavior as threatening, others who saw the vessel engaged in dredging mistook it for a disabled steamer.[25]

The dominant maritime attitude toward naturalists on naval vessels early in the nineteenth century had been one of derision. In the early tradition of naval service, "a philosopher afloat used to be considered as unlucky a shipmate as a cat or a corpse." On the North Pacific Exploring Expedition, William Stimpson encountered this attitude, with profound consequences for his scientific work. Particularly under the arbitrary orders of the first commander, Cadwallader Ringgold, Stimpson chafed at events and decisions that prevented him from collecting as much as or where he liked. First, Ringgold insisted that Stimpson sail with the flagship, which visited only major ports, rather than the surveying brigs, which explored zoologically unknown areas. Ringgold also frequently refused the naturalists permission to land while granting permission to officers. In the months before he was removed from command for mental instability, Ringgold ordered Stimpson to discharge the native boatmen that he had hired at his own expense after being denied use of the ship's boats for dredging. Finally, Ringgold refused to allow "anything to be preserved on board the ship which will make any dirt or create the slightest smell." This order virtually ended Stimpson's work until John Rodgers replaced Ringgold as commander.[26]

Ringgold's eccentric behavior led to extreme problems for Stimpson. But naturalists on naval ships more typically faced less severe though disparaging attitudes on the part of officers and crew toward natural-history work. The expedition botanist, Charles Wright, lamented, "the majority of the [officers'] mess have a most sovereign contempt for science and no esteem for its devotees." When the expedition departed, Stimpson was a twenty-year-old in a responsible position who naively assumed that everyone must be as fascinated by invertebrates as he was. As a result, both officers and men ridiculed his excitement and enthusiasm over seaweed and floating logs rescued from the sea for their animal inhabitants. A decade after the expedition, Stimpson still remembered the sting of taunts he endured, confiding to Dall, who encountered similar problems on an expedition, "I am familiar with the nature of the stumbling blocks which brainless officials delight to cast in the way of men of science. You must keep a stiff upper lip, mind your own business, and resent any interference in it by others unauthorized." Although Stimpson faced outright resentment from his naval shipmates, he felt comfortable in the physical environment of working vessels, having worked on a fishing boat. He manipulated the dredge by himself until almost a year into the expedition, when Ringgold finally assigned seaman Salvador Pelkey to help him. With one exception, he never got seasick. His discomfort on the expedition derived from the cultural clash on the *Vincennes* between his role as scientist and the habits and expectations of naval officers and crew.[27]

In the decades following that expedition, especially on the ships of the U.S. Coast Survey and the British Hydrographic Office, officers became more accustomed to working with scientists at sea. Familiarity mellowed the kind of nastiness that Stimpson had encountered into friendlier jibes. By the 1860s and 1870s, as scientists went to sea on naval vessels in increasing numbers, relations between the two groups became noticeably more restrained and polite. When the mariners laughed at naturalists' odd behavior and preoccupations, they did so more gently than in the past. One young officer teased the naturalists who "paddle and wade about, putting spade-fuls [of mud] into successively finer and finer sieves, till nothing remains but the minute shells." In return, Thomson dubbed the *Challenger's* chief officers "ministers of cleanliness and order." He made a point of recording his "debt of gratitude" to the common sailors on the *Challenger.* Calling them "my friends the blue-jackets," he praised them for treating the scientists "as civilians," and showing them as much respect as they did their officers.[28]

If the officers and common sailors found strange the scientists' habit of covering the deck with mud and mucking about in it, exclaiming over the shapeless, colorless animals, the scientists found naval culture even more foreign. They readily acknowledged their ignorance of the unfamiliar world. Thomson drolly noted that the naval officers referred to the naturalists as "'philosophers' . . .—not I fear from the proper feeling of respect, but rather with good natured indulgence." He admitted that scientific educations were sadly deficient in "the matter of cringles & toggles & grummets & other implements by means of which England holds her place among the nations."[29]

Not all differences of perspective were so light-hearted. Conflict between the goals of scientists and mariners of all levels was deeply rooted in the political culture of ships. Scientists who failed to understand and negotiate the social and political dynamics on board compromised their scientific work. Wallich's voyage on the *Bulldog* in 1860 provides a vivid example of how scientists' shipboard conflicts manifested themselves differently with officers and crew. In the case of social equals, tensions erupted into direct confrontation. Wallich constantly argued with Captain Leopold McClintock, complaining that the crew were not employing sounding devices that retrieved bottom samples frequently enough. After weeks of mutual frustration, McClintock snapped sarcastically, "I suppose you would like to have a diving bell sent down," a ridiculous proposition for working in thousands of fathoms. But no matter what the disagreement the captain always had the final say.[30]

The crew was not as free as the scientists to complain openly. The *Bulldog* sailors relied instead on time-honored forms of protest such as stealing officers' food, deserting, and even taking Wallich's boots to express their dissatisfaction with what must have been an extremely arduous and unpleasant cruise. If desertions are a good index, then Wallich's frequent complaints about the *Bulldog* voyage were well grounded. Three months into the expedition, fourteen men deserted. The next few nights, more men attempted to run away until, as Wallich reported, "the men are evidently in far from a proper state. Today they applied for leave to go ashore in a body!!" He added unnecessarily, "Of course it was refused." That day, all of the ships' boats were hoisted up to prevent more desertions. On the *Challenger,* Joseph Matkin likewise observed instances of the crew stealing food from the officers' mess and deserting the ship, behaviors which signaled their dissatisfaction. The hard work of dredging and sounding, especially on very long voyages such as *Challenger's* three-year trek, frequently drove sailors to jump ship. During the first five months of that voyage, Captain Nares made the tactical decision to land in Halifax instead of New York to discourage the rate of desertion, which he felt would be high in the more attractive port.[31]

The *Bulldog* crew blamed Wallich as much as any of their officers for their hard lot. Besides causing extra work, Wallich condescended to them. He assumed that the sailors would lose valuable bottom sediments in the "sort of scramble to see what was in the apparatus." So he instituted a policy that only he was allowed to extract sediment from the sounding device. In great depths, sounding apparatuses were sent down weighted. The detaching mechanism for the sinkers also triggered the valve that trapped the bottom samples. One day when the apparatus failed and came up empty, with the sinker still attached, the sailors saw their chance to embarrass and annoy Wallich by adhering to the letter of his law. Lugging the 118-pound sinker below to his cabin, they roused him and solemnly, with straight faces, showed him a film of mud on one side. This sample, Wallich tersely recorded, he did not "deem" worth preserving.[32]

Even on voyages not as fraught with conflict as that of the *Bulldog,* scientists and officers often disagreed about the conduct of collecting operations. Before the 1860s, naturalists dredged alone or supervised a small hired crew. In depths over 100 or 200 fathoms, dredging required not only many hands, but also someone skilled at coordinating their work. On naval vessels, the watch officer took over supervision of dredging from scientists. Once steam machinery was introduced to hoist sounding and dredging apparatus, the officer in charge had to orchestrate the crew's labor with the work of the engineers who ran the machinery. Scientists had to stand aside, letting officers decide where and when to dredge and sound, then watch and wait until sailors finally emptied the nets on deck.

Officers' complete control over managing deep-sea operations led, when the devices came up empty, to heated debates over whether or not they had even reached the bottom. The naturalist John Murray archly observed in his journal that the statistics on sounding and dredging posted in the wardroom did not include attempts in which instruments were lost. Joseph

Matkin, the *Challenger*'s steward's assistant, frequently noted in his letters home instances in which the crew lost dredges or other equipment over the side, even in most cases pointing out how much the lost items had cost. When scientists' independent work, such as deploying tow nets, interfered with hydrographic work, "there was a row."[33] Such conflict usually resulted in an order to the scientists to cease their activity.

When scientists lost the control over shipboard operations that they had exercised on hired fishing vessels and private yachts, they retaliated by asserting their authority as scientific experts. They routinely questioned officers' species identifications and generally belittled naval competence in all branches of natural history. In his expedition journal, Stimpson ridiculed claims by several officers that they had seen a penguin in the Coral Sea. With difficulty, he managed to convince the deck officer of the geographical improbability of such a sighting, so that the it would not be recorded in the official log. The next day, Stimpson reported archly in his journal that the "penguin" had been an escaped chicken. Revealing a similar attitude two decades later, *Challenger* naturalist John Murray listed in his journal the bird sightings for one day in which four different officers identified four species of birds, including one species of which they had taken no specimens during the entire expedition. In conclusion Murray noted laconically, "Myself I saw no birds."[34]

Overall, scientists and naval personnel faced each other with vastly different expectations, assumptions, and interests about the meanings of life and work at sea. Issues involving trust and authority had to be worked out. Without scientists or their work to consider, captains, officers, and sailors had traditional, although never static, ways of negotiating with one another and asserting their interests. Captains might withhold grog or dole out extra rations, while sailors might steal from officers or desert. The addition of outsiders to naval and maritime hierarchies added a new element that had to be reckoned with in order to allow gathering scientific knowledge about the ocean. Together, and at sea, scientists and seamen restructured the ship and adapted existing maritime practices and traditions to forge a new maritime culture that had room for the scientific outsiders and their work.

EARLY OCEAN scientists, including the *Challenger* organizers, were cognizant of their role in forging a functional workplace for scientific investigation of the depths. Thomson described the effort to associate independent civilian scientists with a man-of-war as a "critical experiment." Both he and Captain Nares declared that experiment a resounding success. Thomson understood the political importance of adequately crediting naval help. He declared that "all the naval officers, without exception" assisted the scientists "in the most friendly spirit." In the first month of the expedition Captain Nares reported to his boss, Hydrographer George Richards, that "the philosophers are all much charmed by their trip thus far." In spite of the serious problems that plagued these kinds of voyages, the new watchword on scientific vessels after midcentury was politeness. Nares's comment that "everyone here is most civil" bespoke the appearance on naval ships of a new level of civility, behind which lay a new commitment for naval officers to facilitate, even participate in, the scientific work conducted on board.[35]

Before the *Challenger* voyages, scientists who boarded naval vessels did so anticipating conflict. During the planning for the North Pacific Exploring Expedition, Spencer Baird recommended to Commander Ringgold that the naturalists be assigned titles, "some appellation other than naturalist, at least as an official designation. They might be termed members of the scientific corps, computer, assistant astronomer, or whatever else may be considered proper to accomplish this object." Before his voyage, Wallich expected that his status as an independent scientist would cause *Bulldog*'s officers to "look upon [him] with something very much akin to jealousy." Naval authorities were not the only people to balk at the prospect of having scientists on board. The engineers of the Atlantic Telegraph Company were alarmed by rumors that scientists would accompany cable-laying voyages. They expressed their hope that only Company officials would accompany the expedition, since "no good can possibly come of the presence of amateurs, while harm may arise in many ways from their being on board." The orders given the first naval vessels whose primary missions were scientific dredging came with stern warnings to the captains that scientific work was to be given priority. Lieutenant-Commander Edward K. Calver, captain of the *Porcupine,* received an official reminder that William B. Carpenter, Charles Wyville Thomson, and John G. Jeffreys were "eminent in the branches of science which this voyage is intended to investigate." His orders stipulated that he follow the wishes of whichever man was on board to superintend scientific operations, affording the scientists every facility at his disposal.[36]

The taming of scientific maritime culture owes a debt to the large number of yachtsmen who undertook ocean science. With their uniforms, dinner parties, and membership dues, yacht clubs had become the sites of a rarefied version of maritime culture. The physical space and social trappings of yachting shaped scientists' expectations of life at sea. When men accustomed to well-appointed yachts and luxurious steamers faced months, or years, at sea on naval vessels, they set about transforming the environment they found there into surroundings more familiar and hospitable.

Shipboard life, however different from the life scientists left ashore, fostered spaces and practices that bore a striking resemblance to places and habits familiar to their class and sex. To a significant extent, naval officers had already created, within their rigid, hierarchical world, places and times that reminded them of home and of their social status on land. The arrival of scientists aboard naval and surveying ships accelerated this process. Between them, officers and scientists superimposed middle-class settings and habits onto ships. They carved out the dining room, smoking room, drawing room, and study. Shooting, drawing, and writing occupied both work and leisure time, while maritime traditions and superstitions were converted to entertainment. The transformation was prompted by the new standards set for ocean travel by luxury steam liners, but it was reinforced by outsiders such as scientists who came to live aboard naval vessels.

More than any other part of life at sea, dinnertime, as a social institution, occupied a position of importance to sailors and passengers alike. Packet and steamship companies recognized this and vied with each other to offer larger, longer, and ever more elaborate meals to occupy and satisfy their passengers. On naval vessels, dinner was a completely different affair for officers and men, as evidenced by the fact that Stimpson, the naturalist on the North Pacific Exploring Expedition, had to interrupt his dredging excursions away from the ship to return for the staggered mealtimes of the crew at noon and five o'clock and the gentlemen at three o'clock. For officers and scientists, the wardroom mess, complimented by wine that they brought along to augment official rations, offered a respite from the working world on deck. The gentlemen on board enjoyed a credible imitation of a respectable dinner party (when the food was good enough to maintain the fiction), albeit in the by then old-fashioned style of the early nineteenth century, without the company of women. After dinner, they retired to the deck, or in bad weather stayed at the table to smoke and talk. Just as colonial bureaucrats in remote outposts maintained contact with civilization by dressing formally, in European clothing, for dinners of European-style food eaten alone, scientists at sea used familiar routines to make a segment of the maritime world comfortable.[37]

On the first Christmas aboard *Challenger*, one month into the cruise, the expedition artist John J. Wild noted that, despite the rough weather, all of the scientific staff felt well enough to appear at dinner. Perhaps it was their victory over the elements as much as the pleasant company that enabled Wild to report that "the evening passed agreeably." After their first year at sea, Thomson recorded another pleasurable Christmas celebration, again in spite of the weather. "At sea never very comfortable especially in the roaring forties where the motion of the ship very much interferes with the cooking arrangements. Everyone seemed in good spirit however. The men were piped into dinner at 12 o'c by the band & all the messes had their 'plum doughs.' We had a glass of champagne in the ward room in the forenoon and drank the toast that the second year might be as pleasant as the first which showed that the first had not been far astray." As Lieutenant Herbert Swire described it, that toast and the whole Christmas celebration were enlivened by "some dozens of bottles of fine old mountain dew, which did warm our hearts exceedingly." One of the Scottish officers, or any of the three Scottish scientists, Thomson, Murray, or James Buchanan, might have provided the libation. On Christmas day, the band played "The Roast Beef of Old England" to call the bluejackets to dinner, and continued to play at intervals during the evening. The officers had a "grand dinner," after which they "mustered" on deck to smoke, sing, and finish the whiskey. Holidays and special events were occasions for reflection and resolve, especially for scientists who sometimes wondered why they were devoting years away from their careers and families. At the New Year's dinner the next week, Thomson noted the traditional sailors' toast to "Absent friends, Sweethearts, and Wives."[38]

In addition to participating in the established rituals associated with gentlemen's shipboard dining, scientists brought new middle-class customs aboard with them. After a day of active dredging, Stimpson declared, "Never was tea so greatful [sic] to me as now when ones appetite is rendered natural by a day of healthful labor, in the delightful pursuit of the little known denizens of the deep sea." Teatime was hardly a traditional seagoing occasion. The attraction of a nonalcoholic drink during the midday stemmed from the new values of sobriety and hard work embraced by the middle classes ashore, values that not everyone on board willingly embraced. The arrival of upper-middle-class traditions carried with it habits such as teatime that changed the texture of life at sea.[39]

At port stops, curious officials and locals flocked to visit the *Challenger,* drawn by its unusual and widely advertised scientific mission. When the captain entertained ladies and gentlemen aboard ship at luncheons, the *Challenger* resembled a large steam yacht making a day sail around the Isle of Wight. Murray described one such excursion from Fayal, on the Azores, with seven ladies and three gentleman visiting. "When we got out of the lee of the Island it became too rough for the ladies, so we put back and kept in smooth water. The captain gave them lunch and the Band got under way and we had the first dance aboard 'Challenger.'" Other dancing occasions were more raucous, as when the "Commander" and the "Professor" waltzed together at the Bahia Cricket Club, "which shook the room some!" Indeed, entertaining local dignitaries and officials was part of the ambassadorial role that HM ships and captains played all over the world.[40]

Entertainment at sea was also influenced by gentlemanly and middle-class values. Music had long been an important element in maritime culture. Sea chanteys helped motivate the crew and coordinate heavy physical labor shared by many hands, such as raising and lowering sails, yards, anchors, and deep-sea sounders. Illiterate common sailors had always used songs to record information and tell stories. Aside from the boatswain's whistle, though, instruments were uncommon in the forecastle. Officers and midshipmen from upper- or middle-class backgrounds were much more likely than common tars to play, and own, musical instruments. Music played an important role in Lieutenant Swire's life on the *Challenger.* In his first letter home, he lovingly described a "beautiful little instrument in the mess . . . called a melodeon" that he liked to hear Lieutenant Pelham Aldrich play. For his musically literate siblings, he compared it to "a harmonium, but softer and more mellow." Music provided a welcome distraction from the tedium of sounding and dredging, as Swire explained: "I play the fiddle nearly all day, at least when I am not on watch or doing other duties, so that I manage to forget to grumble."[41]

Sailors who faced long periods away from home, particularly those on exploring or whaling vessels, turned to music and other forms of entertainment for diversion. Arctic explorers who wintered on the ice devoted enormous amounts of time and energy to theatrical sketches and other amusements, which sometimes provided the only activity for weeks at a time. Swire frequently mentioned occasions when the *Challenger* band, of which he seems to have been a member, played for the crew, officers, or visitors. On the expedition's last Christmas Eve aboard, he chronicled an elaborate after-dinner extravaganza, including songs, fiddling, recitations, and readings. "These performances," he noted, "have since become a fortnightly institution on board." The addition of a bowl of punch turned such occasions into parties.[42]

Shooting, a popular shipboard and onshore activity, is another segment of maritime life altered by the arrival of significant numbers of gentlemen. Naval officers who liked to shoot often combined their sporting interests with natural-history collecting, which explains why some officers became jealous of full-time naturalists who took away that pastime. Lieutenant Swire executed his surveying duties unwillingly, "having an eye rather to the slaughter of a duck than to dreary angle-taking." On the *Challenger,* the association of sport with collecting is evident in Murray's praise of his friend Lord Campbell, one of the sub-lieutenants, for shooting a black albatross "on the wing." Scientists, too, found it difficult to resist the temptations of sporting pastimes. The young German naturalist Rudolph von Willemoes-Suhm found it "hard to make a compromise between the duties of a scientific zoologist and the desire to shoot a rifle."[43]

Shark fishing was another pastime that provided amusement for sportsmen and spectators alike. *Challenger* expedition members arrived in the area of St. Paul's Rocks in the South Atlantic with visions of pitting themselves against the many sharks they knew to dwell in those waters. They prepared the best bait and hooks to use, gleaning advice from accounts of shark fishing in the area in expedition narratives. In his shipboard journal, Murray described John Hynes, the assistant paymaster, as the most enthusiastic shark fisher. Sub-Lieutenant Andrew Balfour described a Sunday spent in pursuit of sharks: "It was good fun catching the sharks as they gave us good sport hauling them in." Common sailors, who were not invited to join in the gentlemanly pastime of shooting, joined enthusiastically in fishing for sharks. Swire remarked upon the "utmost hatred" that seamen held for the "ravenous monsters," which they "torture[d] in many ingenious ways before finally hacking them to pieces."[44]

While officers and scientists imported new middle-class habits onto ships, they also appropriated and transformed traditional shipboard practices. Following the precedent of captains who doled out additional grog for unusually hard work, Chief Scientist Thomson produced champagne for the officers' mess after the capture of a crinoid new to science. Partially maintaining the traditional separation of the captain from the rest of the crew, Captain Nares and Thomson organized a "fore-cabin club" for card-playing with Thomson's secretary as well as Nares's second in command. Often landlubbers participated more eagerly than the sailors in traditional maritime activities. When the *Challenger* crossed the equator for the first time, Lieutenant Swire noted that he and the other novice officers "may consider that we have been lucky in escaping the levee which Neptune usually holds on these occasions." Traditionally, sailors crossing the line for the first time were initiated by experienced sailors before a kangaroo court presided over by "Neptune." Rough treatment, including shaving and dunking of greenhorns, could sometimes be avoided by paying a fine. This rite of passage brought neophytes into a brotherhood of blue-water

Shark fishing was a popular pastime aboard HMS *Challenger*, as on expeditions in general, one that common sailors as well as officers and scientists enjoyed. (From Charles Wyville Thomson and John Murray, ed., *Report of the Scientific Results of the Exploring Voyage of the HMS "Challenger," 1873–76: Narrative of the Cruise*, vol. 1 [London: HMSO, 1885], 560; courtesy of Dr. David C. Bossard.)

sailors and also relieved social tensions between officers and crew as ships passed through equatorial doldrums. While the *Challenger* sailors did not think that the day's activities fulfilled maritime tradition, the scientists did. John Murray noted that his crossing cost him two quarts of Moselle, no doubt extorted by his social equals, the officers, rather than the crew.[45]

From the 1820s on, collecting folklore about disappearing customs and peoples became a priority for explorers and travelers. Landlubbers on ships were intrigued by maritime customs and entertained by maritime superstitions The offer of tribute to Neptune was often carelessly dispensed with by men who navigated scientifically, using specialized equipment and wind and current charts. Scientists thought sailors' tales of sighting sea serpents and other fantastical creatures were extremely dubious. Finding a true sea serpent, scientists maintained, was highly unlikely, although they did not rule out the existence of monsters of genera other than serpents. They were quite unconvinced of the existence of the more fantastic sea creatures. Wallich marveled at a yarn told by an old Irishman who said he saw a man catch a mermaid off the coast of Newfoundland. The story teller related that the man had had two children with her, fulfilling the common pattern of folkloric stories about a different sort of mythical sea creatures, selkies.

Mermaids were a popular motif for ocean scientists and others who encountered the maritime world at midcentury. (From Charles Wyville Thomson and John Murray, ed., *Report of the Scientific Results of the Exploring Voyage of the HMS "Challenger," 1873–76: Narrative of the Cruise*, vol. 1 [London: HMSO, 1885], xxix; courtesy of Dr. David C. Bossard.)

(Named after the Orkney Island word for seal, selkies are seal people who can shed their skin and come ashore in human form. A man who finds or steals the skin of a female can force her to marry him, although she willl return to the sea if she recovers her skin.) Wallich commented in amazement that the old man had evidently told the story so often that he had come to believe it.[46]

One superstition tested repeatedly on nineteenth-century scientific exploring expeditions was the stricture against killing albatrosses while at sea, made famous by Samuel Taylor Coleridge's "Rime of the Ancient Mariner." Albatrosses being one of the sights landsmen learned to expect at sea, scientists were thrilled by their first glimpses of these birds. William Stimpson mused, "This wonderful bird has a most striking appearance when seen from afar, sailing noiselessly over the ocean seemingly without a motion, or a single vibration of its wide spreading wings. I was never more vividly reminded of a 'spirit brooding over the dark surface of the waves.'" Poetry did not, however, prevent him, a week later, from capturing and killing albatrosses to preserve their skins. One of the expedition's officers

An officer on HMS *Challenger* catching a sea bird using a baited hook and line. Officers and scientists also enjoyed shooting birds. Sailors sometimes expressed concern when scientists violated maritime taboo by killing albatrosses. (From Charles Wyville Thomson and John Murray, ed., *Report of the Scientific Results of the Exploring Voyage of the HMS "Challenger," 1873–76: Narrative of the Cruise*, vol. 1 [London: HMSO, 1885], 848; courtesy of Dr. David C. Bossard.)

expressed remorse at the huge number of birds that he and the other officers had shot one day, not for food or as scientific specimens, but for sport. He recorded that the common sailors feared that a shipwreck would result from the slaughter. On the *Challenger,* only officers and scientists, not crew, killed albatrosses, partly because the upper-class association of shooting precluded their participation, although crew members did collect other sorts of animals, but also because they were unwilling to break the long-standing taboo. Scientists delighted in poking fun at such superstitions, as when Murray noted sardonically, "Have been sailing all day. Not an Albatross has been seen today, so that the first day we really got a good trade wind, the Albatross left us."[47]

Of all the cultural transformations on board midcentury vessels, reading, writing, and drawing most clearly reflected the influence on sea life of middle-class values about work and leisure. In addition to filling leisure time at sea, all were important aspects of both scientific and surveying work and therefore they were important activities in the construction of the new maritime culture that facilitated scientific research at sea. For scientific gentlemen, retreating behind books and microscopes was an effective means of escaping from the maritime world. Their work defined their role aboard ships and represented an important way to distinguish themselves as gentlemen and scientists. In aspects of their work that resembled what they did ashore—studying, writing, reading, microscopy, dissecting, preserving—they kept up their regular activities, merely adapting them to more cramped, less stable areas. Their work often separated them physically from the officers and crew. Huxley described his daily routine this way: "Shut up as I am in the midst of this busy world, I manage to lead more completely than I have ever done, perhaps, the solitary life of the student."[48]

Books were an important part of maritime culture, and most ships carried a library. But scientists required larger libraries than those usually carried on ships, even exploring vessels, because they needed monographs on species they expected to encounter in addition to the usual geographies of regions they planned to visit. The North Pacific Exploring Expedition carried over a thousand books. When Wallich sought Joseph Hooker's advice before the *Bulldog* expedition, he asked which books he would need and where he could get them. The *Challenger* had an enormous library, with books stowed in the naturalists' workroom, the fore- and after-cabins, the wardroom, and anywhere else they would fit. In addition to books officially procured by the Admiralty and the Hydrographic Office and those owned by expedition members, Macmillan, which published all of Thomson's books, donated "a case of about 50 volumes of his newest publications." Lieutenant Swire, who was not a scientist, could assure his family that they were "not likely to suffer from a book famine." Even so, the zoologists sometimes found the library lacking in specialized literature on some faunal groups.[49]

Writing had always been part of maritime work through log-keeping. On a personal level, correspondence kept all mariners, including scientists, connected to the people they had left behind, as when Huxley wrote to his fiancée from HMS *Rattlesnake,* "The excitement of letter reading and writing has made the fortnight we have passed here go by like a dream." Letter-writing also played an important part in maintaining professional contact between naturalists. Dall wrote to Jeffreys, with genuine gratitude, from his ship in Alaskan waters, "Please accept my thanks for [your letter] and believe that I fully appreciate your kindness in writing to alleviate the isolation in which we have passed the last 13 months." For a scientist who imagined himself as the next Huxley or Charles Darwin, journal keeping not only recorded observations and collections, but also strengthened his sense of being a scientific explorer. Writing could likewise serve as a way for sailors "before the mast" to distinguish themselves from their peers in much the same way that scientists' dissecting and microscopic work distinguished them from their naval colleagues. Joseph Matkin, the steward's assistant on the *Challenger*, spent hours composing long letters to his family about his experiences. As the son of a printer, Matkin no doubt had middle-class ambitions that may have inspired him to stay at the cramped desk he used to keep the steward's records to do his own writing, eschewing the company of fellow crew members with whom he felt he had little in common.[50]

Like writing, drawing and painting was an integral component of both hydrographic and scientific work. The *Challenger* sailed with an official artist, John J. Wild, whose job included making scientific drawings of delicate marine invertebrates and deep-water fishes. He also recorded events on the expedition, including port stops and encounters with native peoples and their material culture. Naturalists did not rely solely on official expedition artists to draw their specimens, because drawing was a central part of natural-history practice. Naturalists' training included instruction in the "grammar" of their visual language so that all naturalists could recognize scientifically important details in well-executed drawings of specimens. Drawing was an important part of surveying work as well, because navigators relied on accurate pictorial representations of shorelines and rocks that were published with charts.[51]

In addition to drawing and painting for work purposes, midcentury officers, sailors, scientists, and even passengers took up their brushes and pencils for pleasure. Naval officers sometimes sketched specimens for scientific use, as when Lieutenant-Commander Joseph Dayman reproduced the microscopic view of Atlantic bottom samples collected by his ships' surgeon. On the *Bulldog*, Wallich described the "regular iceberg mania" that seized the wardroom in the northern latitudes, inspiring officers to try to capture pictorially for "the enlightenment of absent friends the wonders of these frozen mountains." Swire also recorded his intention to try to depict with his paintbrush "the huge icebergs, the interminable pack, the marvelous sky, and the beautiful birds which fly in it." He also noted, with evident pride, that several of his india-ink shoreline sketches for charts had

A watercolor by the HMS *Challenger* cooper, Benjamin Shephard, "HMS Challenger Sailing amongst the Ice." Mariners seeing icebergs for the first time routinely tried to capture the sight with pencils or paint. (Courtesy of J. Welles Henderson, personal collection.)

been sent home to be engraved. Lieutenant Aldrich's carefully wrought *Challenger* journal, with beautiful handwriting and detailed watercolor sketches, testifies to the importance that visual aesthetics held for him.[52]

The production of art was by no means limited to officers and scientists; maritime art was often produced by common sailors. Whalers, for example, made scrimshaw, or engravings on pieces of whale bones or teeth. A sketchbook of watercolors by a cooper aboard the *Challenger* offers a glimpse of a common sailor engaged in a particularly upper- and middle-class art form. Benjamin Shephard entered naval service as a cooper in 1862, deserted five years later, but returned, and served for the entire *Challenger* voyage. The sketchbook contains thirty-four watercolors covering only the first third of the expedition, which suggests that other books probably existed. His ship renditions were extremely accurate, a common feature of maritime art. Like many "true men of sail," however, he frequently refused to depict the smokestack. Most of his paintings show ports, but a series painted in the Antarctic reveals the same interest in icebergs manifested by his fellow artists. Similarities between his paintings and Wild's, especially their depictions of the famous technical accomplishment of making the *Challenger* fast to St. Paul's Rocks, indicate that they either worked side by side, copied each other, or both worked from photographs. Like Joseph Matkin, Shephard probably relied on his painting to define himself in relation to his peers.[53]

Activities such as painting, journal keeping, listening to and performing music, smoking, and dining with friends helped voyage time to pass pleasantly for scientists. But most of them nevertheless enthusiastically welcomed any chance to get off the ship. Stops in small ports or at relatively unknown islands offered valuable collection opportunities on the shore, often referred to with the gentleman-naturalist term "rambles." Longer port stops, especially for ship repairs, gave scientists welcome opportunities to move workspace ashore for a time. Alternatively, because scientists, unlike naval officers, had no official ship duties during port calls, they sometimes used these occasions to make inland trips, as when Murray and some other *Challenger* scientists left Halifax on May 12, 1874, to travel to Boston, London and Winsor, Ontario, Chicago, Louisiana, Niagara Falls, Albany, and New York City before sailing to Bermuda to meet the *Challenger* there eighteen days later. When they made landfall in less sparsely settled parts of the British Empire, they settled for challenging local cricket clubs to matches.[54]

Scientists were not always intent on retreating from ship life. Some went to sea enthusiastically, primed by experiences on fishing vessels or yachts and excited by the image of themselves as heroic explorers, pitting themselves against the ocean. Others grew familiar with maritime culture only after they boarded ships, but came to appreciate aspects of it. Huxley admitted that although he was not by nature a methodical person, his daily routine aboard the *Rattlesnake* had the "clockwork regularity" of naval life. Moseley welcomed the opportunity to take on extra duties in the Magnetic Department, work that he began as "a perfect stranger to it." Soon he was as proficient as his naval colleagues at making the observations, though he confessed that he continued to find some of the calculations "a little strange." By the end of a season of Coast Survey field work, Dall had become knowledgeable about the characteristics and technical specifications of ships, equipment, and techniques needed to conduct deep water hydrography. He was so committed to the work that he stayed through two extra seasons in spite of urgings from Baird and other East Coast naturalist friends to return to Washington.[55]

Shared work and entertainment fostered enduring friendships between scientists and sailors. Some scientists expressed sorrow at leaving their adopted homes and shipboard life when the time came. Even George Wallich, after months of complaining, admitted on the last day of the voyage, "Now that the cruise is all but over, I feel I leave friends and friendships among them." Both Huxley and Dall corresponded with former shipmates who were common sailors. After the *Challenger* voyage, Swire complimented one of his "greatest friends," Moseley, for his "imperturbable good humour, which made him absolutely proof against all the shafts with which naval wit was never tired of trying the mettle of those whom we called our philosophers, and which enabled him at last to completely turn the tables on his funny friends."[56]

But neither the formation of friendships nor the successful completion of voyages entirely eradicated friction between scientists and sailors. Naval officers maintained, "Somehow naval men always get along together better than other people." Problems often arose out of incongruent work rhythms, as when science and a ship's housekeeping conflicted. Huxley warned prospective seagoing naturalists that a net deployed overboard required constant supervision or "it will be pretty certainly encumbered with the products of the 'Head' & other rubbish." Landlubbers on ships tended to retain a landsman's perspective, sometimes consciously, sometimes not. Where sailors gazing over the rail saw jellyfish, scientists observed "many *specimens* . . . float[ing] past the ship." And although common sailors helped with scientific collecting, they saw animals in starkly economic terms, as commodities rather than intellectual, national resources. Even though members of the Wilkes Expedition were forbidden to make personal collections, Charles Erskine sold an orange-colored cowrie shell and a tortoise shell to an English naturalist rather than surrender them to the expedition's scientific staff. To the great irritation of naval officers, the *Challenger* naturalists "persist[ed] in calling things by either shore-going names, or terms which they have picked up in merchant passenger steamers." In the middle of the *Challenger* expedition, Thomson still reckoned time in the land unit of university sessions. Sailors, by contrast, kept their lives organized around "watches" even during overland explorations.[57]

Yet despite their differences, for the most part scientists and sailors learned to live comfortably with each other and work together effectively. Watching one another perform similar kinds of work, such as reading, writing, and drawing, encouraged scientists and officers to view each other as colleagues. Most of all, sharing living space built a social bridge to the wardroom that facilitated collecting and encouraged communication between landlubber naturalists and hydrographers trained in the physical sciences. To a certain extent, scientists succeeded in importing their middle-class habits into the small worlds of the ships they boarded. The new, more genteel form of maritime culture that resulted from this contact contributed to the formation of a work culture for the emerging discipline of oceanography.

Epilogue

Roll on, thou deep and dark blue Ocean—roll!
Ten thousand fleets sweep over thee in vain;
Man marks the earth with ruin—his control
Stops with the shore;—upon the watery plain
The wrecks are all thy deed, nor doth remain
A shadow of man's ravage . . .

—George Gordon, Lord Byron, *Childe Harolde's
Pilgrimage,* 1818

Environmentalists in recent years have raised alarms about the fragility of the seas. The burgeoning sense of an environmental crisis of the ocean is emerging hand in hand with the dawning recognition that our everyday lives affect, and are influenced by, the ocean. The idea that terrestrial environments have complicated histories tied to human history is not new. Yet our perception of the ocean remains quite distinct. Almost two centuries ago, Lord Byron evoked the ocean's eternity, its indifference to human existence. Today we still understand the ocean as "empty of history, utterly without a past."[1] If we perceive the ocean as eternal and impervious to human actions, then it becomes impossible to make sense of the current environmental crisis. How can there be an urgent crisis if the human relationship with the sea has been static?

Yet the ocean does have a history. Coastal residents have of course always lived with the bounty and tragedy of the sea, but the interdependence between the oceans and every person on earth tightened perceptibly, even dramatically, in the mid-nineteenth century. The deep ocean is a realm with an identifiable, historical relationship to human activity, one that began in the era of mid-nineteenth-century imperialism and industrialization and has intensified with time. The midcentury discovery of the ocean's depths set precedents for resource use that continue today; nowhere but on the sea are we still primarily hunters rather than farmers. Although many sentences of Herman Melville and Richard Henry Dana ring true to sailors today, the ocean is no more immune to the march of time than other environments in which humans work, play, explore, and reflect.

Our current understanding of the ocean environment, though it began with the working knowledge of mariners, whalers, and explorers, owes much to the investigations and inventions of the mid-nineteenth century. Before then, the ocean—as a place—was not an object of scientific study. True, individual natural philosophers investigated the characteristics of seawater, and people who relied on knowledge of the sea for their livelihood and lives paid attention to winds, currents, fish, and whales. But the process by which the ocean attracted scientific attention required a shift away from understanding and experiencing the ocean as an expansive divide, a watery highway, or an unfathomable barrier between places. It involved a cultural redefinition of the sea as a destination and a location with new meaning for the Western world.

The history of science and technology provides an essential lens through which to view the ocean's history. British and American scientists and hydrographers began to study the ocean's depths between about 1840 and 1880, the period when cultural interest in the deep ocean peaked in both countries. Scientific analysis of the sea and its denizens coincided with a widespread economic and social awakening to the maritime world, and scientists contributed as much to the construction of the ocean environment as to comprehension of the depths.

The United States and Great Britain initiated the earliest sustained efforts to understand the depths of the sea, although scientists from other countries, especially Scandinavia, certainly contributed to early ocean science. The two nations shared many maritime traditions and had common strategic naval aims. They became literally connected by the Atlantic cable. Their scientists were members of overlapping intellectual circles and their sailors moved from one country's ships to the other's, linking their navies and merchant fleets. Both were conscious of their geographic separation and freely acted upon their considerable expansionist impulses. Britain defined itself predominantly as an island with a need to master the sea around it, and the United States maintained a strong maritime orientation even during decades of vigorous westward land expansion. Both countries strongly supported hydrographic surveying institutions that created the world's sea charts. Both for many years continued to send civilian naturalists on military voyages of exploration and survey missions, a habit that France, for example, had abandoned by 1800. Their scientific and technical communities shared knowledge about the ocean and technical information about how to acquire it.[2]

Scientific interest in the sea intersected with commercial interests, inspired by shipping and sperm whaling, as well as political interest, mobilized by submarine telegraphy, to define the ocean as important new territory. The whaling industry boomed in the 1820s and 1830s, providing tantalizing reports from waters never previously traversed. A tradition of scientific, imperialistic exploration, exemplified by Charles Darwin's *Beagle* voyage, Thomas Huxley's on *Rattlesnake,* and the United States Exploring Expeditions, remained a vital part of serious natural history and an essential ingredient in the foreign policy of these premier maritime nations. Efforts to bridge the European and North American continents with a trans-Atlantic telegraph cable reflected the brash confidence of Victorian engineers and entrepreneurs.

The Anglo-American discovery of the deep ocean was as cultural as it was political and scientific. Popular awareness of the blue water beyond the shore increased around midcentury, partly because of the popularity of marine natural history and the publicity surrounding the Atlantic submarine cable-laying attempts. The social currency of seaside holidays, ocean travel, and yachting drew naturalists to the sea. The ocean entered everyday life in the form of aquaria, sailor suits, and maritime novels. Written by the first professional writers to experience work at sea, these books joined travel narratives of ocean voyages as vehicles that fed upon and promoted popular interest in the ocean. Scientists, as carried away by popular culture as their contemporaries, swiftly made the ocean a focus of their attention.

As the sea, including its greatest depths, acquired economic, political, and social importance, Britain and the United States undertook to discover, name, and chart its bottom contours, species, and water masses. These endeavors began at midcentury, when most of the globe had been discovered, named, and claimed. Only places such as the interior of Africa, where malaria effectively kept Westerners out, remained more inaccessible than the depths until near the end of the nineteenth century. Scientific exploration of the deep sea, as well as that of the atmosphere and the Arctic regions, reflected the desire to tackle new frontiers, to extend the realm of imperialist expansion.

The deep sea yielded quickly to the Western propensity for reshaping and redefining the natural environment for human use, most explicitly as a safe home for submarine telegraph cables. The potential for trans-Atlantic telegraphy bolstered nationalist motives for ocean exploration. Contemporary attitudes toward land and terrestrial natural resources transferred easily to the sea, so that the ocean floor became unclaimed territory, "almost designed by Providence," to protect submarine cables.[3] Explorers and scientists fully expected the national use of oceanic resources, such as whales, fish, and minerals, to follow closely upon their pioneering work.

By the 1860s, the ocean was recognized as a promising research site not just by amateur, beachcombing naturalists, but by professional zoologists and, increasingly, by physical scientists as well. As scientific interest in the deep ocean broadened, biological and physical scientists banded together to argue for national resources to support major investigations. This group mobilized the support of the Admiralty and the Royal Society for a series of cruises to dredge and study the deep sea. The most famous of these, the HMS *Challenger* expedition, sailed in 1872 for a four-year circumnavigation voyage devoted to studying the world's oceans. Many scientists believe that the *Challenger* expedition initiated oceanography. Certainly its fifty-volume report formed the foundation for the new discipline. But the *Challenger* voyage itself involved practices, motivations, equipment, and questions stemming from a wide range of nineteenth-century maritime and terrestrial activities.

Beginning in the 1860s and 1870s, aboard *Challenger* and other vessels, oceanography began to develop in the crucible of oceangoing ships, on whose decks landlubber naturalists faced the challenge of integrating their work into the ships' physical and social structures. Scientists joined the swelling ranks of middle-class seagoers who eagerly explored the unfamiliar maritime world. Together with officers who were their social equals, scientists created a safer and tamer version of maritime culture, one that not only made them comfortable in the alien world of oceangoing ships, but also promoted their scientific work.

As oceanography became the purview of professional scientists, the post-*Challenger* generation of scientists studied the sea in a more business-like, less romantic manner. Instead of voyage narratives, they wrote textbooks, such as *The Ocean* by John Murray and Johan Hjort. The story of the discovery of the major mid-ocean mountain chain, the Mid-Atlantic Ridge, did not prompt front-page newspaper coverage, as had the submarine telegraph exploits of the 1850s and 1860s. Analysis of the *Lightning* and *Porcupine* results had convinced Lieutenant T. H. Tizard and chief scientist Charles Wyville Thomson of the existence of a submarine ridge across the channel between Scotland and the Faroe banks. In 1880 Tizard and Murray, working from HMS *Knight Errant*, located the ridge but were unable to compare marine life on both sides because of insufficient trawling gear. Two years later they tried again, with success, from the newly launched *Triton*, the only British vessel built in the nineteenth century especially for surveying work. Results from these voyages appeared in the scientific literature, most prominently in the pages of *Nature*, but did not provide grist for adventure stories.[4]

By that time, scientific interest in the ocean's great depths had begun to attenuate. Some scientists remained interested in the deepest mysteries of the sea, of course. But many marine-oriented scientists concentrated on coastal and inshore waters and on fisheries-related research. The historian Margaret Deacon has noted that scientific curiosity about the oceans has waxed and waned several times since ancient times, including a period of abeyance after 1900. Writers at that time sounded the death knell of the "age of sail." In his 1909 novel *Tono-Bungay* H. G. Wells insisted, "There is no romance about the sea in a small sailing ship as I saw it. The romance is in the mind of the landsman dreamer."[5] Cultural interest in the ocean's depths had likewise declined. By the 1880s, submarine telegraph cables no longer evoked curiosity about the depths; they had instead become symbols of the geographic extension of national empires.[6] Interest in maritime novels persisted, but Joseph Conrad's stories explored primarily the psychological dimensions of the sea whereas Herman Melville and Richard H. Dana had done that and also aimed to portray shipboard life realistically. The midcentury home aquarium fad lasted only briefly, followed by a similarly widespread but short-lived public aquarium craze. Interest in popular natural history of the sea persisted longer, but without the intensity characteristic of the mid-Victorian era. With the advent of the automobile, inland vacation spots not accessible by rail competed with the seashore as preferred vacation sites. The ocean's depths lost their firm hold on the Anglo-American cultural imagination. By the end of the nineteenth century, the sea floor, and increasingly the intermediate waters, aroused the interest of only a handful of oceanographers.

Yet the seeds had been sown for the next cycles of fascination with the sea. As John Gwyn Jeffreys and William Carpenter put it, all the "grand work" of the *Lightning,* the *Porcupine,* and even the *Challenger* had only "effected . . . to scrape in an imperfect manner the surface of a few scores of acres." Not even if every civilized nation sent out similar expeditions every year for the next century, they asserted, would this vast field of study be exhausted. The decade of the 1960s perhaps most strongly resembled the mid-nineteenth century in the confluence of intense public fascination with, and scientific interest in, the sea. Roger Revelle, of the Scripps Institution of Oceanography, described the postwar era of oceanography as "the beginning of our great age of exploration, which I think was one of the greatest periods of exploration of the earth, . . . that compares in many ways with the exploration of the fifteenth and sixteenth centuries."[7]

Again today there is renewed interest in "exploration" for its own sake, especially of the deep ocean. A U.S. presidential panel in 2000 recommended that the government spend $75 million a year on open-ended exploration of the ocean as a counterweight to highly specialized and tightly focused, hypothesis-driven research. The National Oceanic and Atmospheric Administration has accordingly opened an Office of Exploration, while the United Nations Intergovernmental Oceanographic Commission discusses equity issues among nations and the potential legal pitfalls of exploration in the waters of coastal nations. Scientists gaze seaward with the conviction, as one put it, that "there's always something out there that's interesting."[8]

Oceanography today retains the imprint of its origins. It remains a science that focuses the effort of physicists, chemists, geologists, biologists, and engineers on the project of understanding a geographic area. Its practitioners do not share a common set of intellectual questions so much as a common definition of themselves as scientists who go to sea.[9] Most of all, oceanography remains a science whose relationship to political and cultural interests continues to be dynamic.[10]

Notes ⚓ *Acknowledgments* ⚓ *Index*

Notes

Abbreviations

ACR	Andrew C. Ramsey Papers, Imperial College Archives, London
ALD	Alder-Norman Correspondence, Natural History Museum, London
APS	American Philosophical Society
BAAS	British Association for the Advancement of Science
CUL	Cambridge University Library
C&WA	Cable and Wireless Archives, London
EUL	Edinburgh University Library, Special Collections Department
HMSO	His [or Her] Majesty's Stationery Office
HO	Hydrographic Office, Taunton, England
ICA	Imperial College Archives, London
IEE	Institution of Electrical Engineers, London
JHP	Joseph Henry Papers
KEW	Royal Botanic Gardens Archives, Kew
MCZ	Museum of Comparative Zoology, Harvard University
NA	National Archives, Washington, D.C.
NHM	Natural History Museum, London
NPEE	North Pacific Exploring Expedition
RGS	Royal Geographical Society
RMS	Royal Microscopical Society
RSA	Royal Society Archives, London
RU	Record Unit
SI, DPH	Smithsonian Institution, Division of Political History, Washington, D.C.
SIA	Smithsonian Institution Archives, Washington D.C.
SFB	Spencer Fullerton Baird Papers
THH	Thomas Henry Huxley Papers
WAL	George C. Wallich Papers, Natural History Museum, London
WHD	William Healey Dall Papers
WHOI	Woods Hole Oceanographic Institution

1. Fathoming the Fathomless

1. Matthew Fontaine Maury, *The Physical Geography of the Sea* (New York: Harper & Brothers, 1855), v.

2. Philip E. Steinberg, *The Social Construction of the Ocean* (Cambridge: Cambridge University Press, 2001), 68–158.

3. British shipping increased in tonnage by 180 percent between 1840 and 1870. Eric J. Evans, *The Forging of the Modern State: Early Industrial Britian, 1783–1870* (London: Longman, 1983); Roy Porter, *London: A Social History* (London: Penguin, 2000).

4. Daniel R. Headrick, *The Invisible Weapon: Telecommunications and International Politics, 1851–1945* (New York: Oxford University Press, 1991).

5. Benjamin W. Labaree, "The Atlantic Paradox," in Labaree, ed., *The Atlantic World of Robert G. Albion* (Middletown, CT: Wesleyan University Press, 1975), 195–217; Robert Foulke, "The Literature of Voyaging," in Patricia Ann Carlson, ed., *Literature and Lore of the Sea*(Amsterdam: Rodolphi, 1986), 1–13.

6. Alain Corbin, *The Lure of the Sea: The Discovery of the Seaside in the Western World, 1750–1840,* trans. Jocelyn Phelps (Cambridge: Polity Press, 1994), 250–281.

7. Steinberg, *Social Construction of the Ocean,* 118–120. Corbin, *Lure of the Sea,* 19–56.

8. Corbin, *Lure of the Sea,* 97–120; Martin J. S. Rudwick, *Scenes from Deep Time: Early Pictorial Representations of the Prehistoric World* (Chicago: University of Chicago Press, 1992).

9. Corbin, *Lure of the Sea,* 57–96.

10. Corbin, *Lure of the Sea,* 250–281; John Stilgoe, *Alongshore* (New Haven: Yale University Press, 1994), 295–367; Lena Lenček and Gideon Bosker, *The Beach: The History of Paradise on Earth* (New York: Viking, 1998).

11. Marcus Redicker, *Between the Devil and the Deep Blue Sea: Merchant Seamen, Pirates, and the Anglo-American World, 1700–1750* (Cambridge: Cambridge University Press, 1987), 155–156; Billy J. Smith, *The "Lower Sort": Philadelphia's Laboring People, 1750–1800* (Ithaca: Cornell University Press, 1990), 155.

12. John Malcolm Brinnin, *The Sway of the Grand Saloon: A Social History of the North Atlantic* (New York: Delacorte Press, 1971), 174.

13. Basil Greenhill and Ann Gifford, *Traveling by Sea in the Nineteenth Century: Interior Design in Victorian Passenger Ships* (London: Adam & Charles Black, 1972); Brinnin, *Sway of the Grand Saloon.*

14. Brinnin, *Sway of the Grand Saloon,* 245; Greenhill and Gifford, *Traveling by Sea,* 45.

15. Brinnin, *Sway of the Grand Saloon,* 153–294.

16. Corbin, *Lure of the Sea,* 234–249; Greenhill and Gifford, *Traveling by Sea,* 53–54; Brinnin, *Sway of the Grand Saloon,* 287.

17. Greenhill and Gifford, *Traveling by Sea,* 11–21; Brinnin, *Sway of the Grand Saloon,* 3–24, 239–248.

18. Charles Bright, *The Story of the Atlantic Cable* (New York: D. Appleton and Co., 1903), 30.

19. Headrick, *Invisible Weapon,* 6–7; David E. Allen, *The Naturalist in Britain* (Princeton: Princeton University Press, 1994, 1976).

20. Cyrus W. Field to William Thomson, June 29, 1859, CUL, Thomson Papers, Add 7342, A 111; Bernard Finn, *Submarine Telegraphy: The Grand Victorian Technology* (London: Science Museum, 1973), 21. The watch fob is in the Judson Collection, Division of Political History, National Museum of American History, Smithsonian Institution. The advertisement is in *Harper's Weekly,* Oct. 16, 1858, p. 671.

21. Winfield M. Thomson and Thomas W. Lawson, *The Lawson History of the America's Cup: A Record of Fifty Years* (Boston: published privately by Thomas W. Lawson, 1902); J. D. Jerrold Kelley, *American Yachts: Their Clubs and Races* (New York: C. Scribner's Sons, 1884); Derek Birley, *Sport and the Making of Britain* (Manchester: Manchester University Press, 1993), 185.

22. C. M. Gavin, *Royal Yachts* (London: Rich & Cowan, 1932), 127–128; François Boucher, *20,000 Years of Fashion: The History of Costume and Personal Adornment* (New York: Harry N. Abrams, 1965), 378, 402.

23. George Brown Goode to Spencer F. Baird, 1878m, SIA, SFB, RU 7002, Box 21. *Chaenopsetta ocellaris* was one name given to the summer flounder. The modern classification of *G. cynoglossus* could not be determined.

24. Foulke, "The Literature of Voyaging," 1–13. Cynthia Fausler Behrman, *Victorian Myths of the Sea* (Athens: Ohio University Press, 1977), 27.

25. "Requisition Book for the U.S. Ship 'Lexington,' 1826–30," Independence Seaport Museum, Philadelphia, Accession no. 69.120.11.

26. W. W., Logbook from the USS. *United States,* 1829–30, Independence Seaport Museum, Accession no. 89.52.2.

27. Thomas H. Huxley to Henrietta Heathorn, Jan. 3, 1849, ICA, THH, Correspondence with Henrietta Heathorn, 41.

28. Margaret S. Creighton, *Rites and Passages: The Experience of American Whaling, 1830–1870* (Cambridge: Cambridge University Press, 1995), 6–15.

29. Corbin, *Lure of the Sea,* 32–56.

30. Charles Darwin, *The Voyage of the "Beagle"* (New York: Mentor, 1972), 177; O. H. K. Spate, "Seamen and Scientists: The Literature of the Pacific, 1697–1798," in Roy MacLeod and Philip F. Rehbock, ed., *Nature in Its Greatest Extent: Western Science in the Pacific* (Honolulu: University of Hawaii Press, 1988), 13–44.

31. Carlson, ed., *Literature and Lore of the Sea;* Behrman, *Victorian Myths of the Sea;* Foulke, "The Literature of Voyaging," 2–5; Robert Foulke, *The Sea Voyage Narrative* (New York: Twayne Publishers, 1997), 1–26.

32. Dennis Berthold, "Cape Horn Passages: Literary Conventions and Nautical Realities," in Carlson, ed., *Literature and Lore of the Sea,* 45; Foulke, "The Literature of Voyaging," 1–13; Foulke, *Sea Voyage Narrative,* 1–26. Steinberg similarly argues that the ocean in the nineteenth century was constructed as a void, an area of anticivilization against which civilized people pitted themselves; see *Social Construction of the Ocean,* 110–158.

33. Behrman, *Victorian Myths of the Sea;* Harold D. Langley, *Social Reform in the US Navy, 1798–1862* (Urbana: University of Illinois Press, 1967).

34. Richard Henry Dana, Jr., *Two Years before the Mast: A Personal Narrative of Life at Sea* (New York: World Publishing Co., 1946), 308–319.

35. Herman Melville, *White-Jacket* (New York: Oxford University Press, 1990), 101–102.

36. Berthold, "Cape Horn Passages," 43–44.

37. For examples of the study of literature and science by historians of science, see Steven Shapin, "Pump and Circumstance: Robert Boyle's Literary Technology," *Social Studies of Science* 14 (1984): 481–520; Peter Dear, "*Totius in verba*: Rhetoric and Authority in the Early Royal Society," *Isis* 76 (1985): 145–161; Rosalind Williams, *Notes on the Underground: An Essay on Technology, Society, and the Imagination* (Cambridge: MIT Press, 1992). The publication since 1993 of *Configurations: A Journal of Literature, Science, and Technology* provides a clear organizational manifestation of this inquiry.

38. Charles Erskine, *Twenty Years before the Mast* (Washington, D.C.: Smithsonian Institution Press, 1985; 1890), 310 (quote), 3–11.

39. Samuel Phillip Lee, *Report and Charts of the Cruise of the US Brig Dolphin* (Washington, D.C.: Beverley Tucker, Printer to the Senate, 1854), Senate Documents, 33rd Cong., 1st Sess., No. 59; Dudley Taylor Cornish and Virginia Jeans Laas, *Lincoln's Lee: The Life of Samuel Phillips Lee, United States Navy, 1812–1897* (Lawrence: University of Kansas Press, 1986), 73–84.

40. Lee, *Cruise of the US Brig Dolphin*, iv.

41. John Mullaly, *The Laying of the Cable, or The Ocean Telegraph; Being a Complete and Authentic Narrative of the Attempt to Lay the Cable across the Entrance to the Gulf of St. Lawrence in 1855, and of the Three Atlantic Telegraph Expeditions of 1857 and 1858* (London: D. Appleton and Co., 1858); Bright, *Story of the Atlantic Cable;* Edward Brailsford Bright and Charles Bright, *The Life Story of the Late Sir Charles Tilston Bright* (Westminster: Archibald Constable and Co., 1899).

42. Edward Frederick Knight, *The Cruise of the "Falcon"* (London, 1880) and *"Falcon" on the Baltic* (London, 1886); Anne Brassey, *Around the World in the Yacht "Sunbeam"* (New York: Henry Holt, 1878); Joshua Slocum, *Sailing around the World Alone* (New York: The Century Co., 1900); John Gwyn Jeffreys, "On the Marine Testacea of the Piedmontese Coast," *Annals and Magazine of Natural History*, 2nd ser., 2, no. 17 (1856): 155–188; Robert MacAndrew, "An Account of Some Zoological Researches Made in British Seas during the Last Summer," *Proceedings of the Literary and Philosophical Society of Liverpool* 1 (1844–45): 90; Richard MacAndrew, "Robert MacAndrew, 1802–1873" (n.p.: privately published, n.d.).

43. Gillian Beer, *Darwin's Plots: Evolutionary Narrative in Darwin, George Eliot, and Nineteenth-Century Fiction* (London: Routledge & Kegan Paul, 1983), 6; Honore Forster, "British Whaling Surgeons in the South Seas, 1823–1843," *The Mariner's Mirror* 74, no. 4 (1988): 401–415; Spate, "Seamen and Scientists," in MacLeod and Rehbock, ed., *Nature in Its Greatest Extent*, 23; Daniel Francis, *A History of World Whaling* (New York: Viking, 1990), 85. Turner also spent a month on a merchant ship, much as writers such as Dana did; Josef W. Konvitz, "Changing Concepts of the Sea, 1550–1950: An Urban Perspective," in Mary Sears and Daniel Merriman, ed., *Oceanography: The Past* (New York: Springer-Verlag, 1980), 32–41.

44. Maury, *Physical Geography of the Sea*.

45. George C. Wallich, *The North Atlantic Sea-Bed* (London: John van Voorst, 1862), 110.

46. Peter H. Kylstra and Arend Meerburg, "Jules Verne, Maury, and the Ocean," *Proceedings of the Royal Society of Edinburgh* B 72, no. 25 (1972): 243–251.

47. Charles Wyville Thomson, *The Voyage of the "Challenger": The Atlantic*, 2 vols. (London: Macmillan, 1877), vol 1., 290–347, 384–388. Pelham Aldrich, [HMS *Challenger* journal], RGS, Library MSS AR 114A, 20.

48. Henry M. Moseley, *Notes by a Naturalist: An Account of Observations Made during the Voyage of H.M.S. "Challenger" round the World in the Years 1872–1876* (London: John Murray, 1892), x; W. J. J. Spry, *The Cruise of the "Challenger": Voyages over Many Seas, Scenes in Many Lands* (New York: Harper & Brothers, 1877), 238–239; Charles Wyville Thomson to [son], [n.d.], Sir Charles Wyville Thomson Papers, MC 17, WHOI Archives; Thomson, *Voyage*, vol. 1, 151–185; Andrew F. Balfour, "HMS Challenger," [journal], vol. 1, Oct. 15, 1873, RGS, Library MSS 113B.

49. Thomson, *Voyage*, vol. 2, 100. Aldrich, [HMS *Challenger* journal].

50. Maury, *Physical Geography of the Sea*, 42, 181, 202 (quote). Eric L. Mills has also noted the frequent comparisons of ocean science to astronomy in "Exploring a Space for Science: The Marine Laboratory as Observatory," *Actas VII Congreso de la Sociedad Española de Historia de las Ciencias y de las Ténicas* 1 (2001): 51–57.

51. John Gwyn Jeffreys, *British Conchology*, vol. 1 (London: John van Voorst, 1862), vi; Moseley, *Notes by a Naturalist*, 517–518; William B. Carpenter to George G. Stokes, June 28, 1871, RSA, MC 9.227.

52. Edward Forbes, *The Natural History of the European Seas*, ed. and continued by Robert Godwin-Austen (London: John van Voorst, 1859), 34; William B. Carpenter, "On Temperature and Animal Life in the Deep Sea, Part III," *Nature* 1 (1870): 563–566; James Glaisher, *Travels in the Air* (London: Richard Bentley, 1871). See also Jennifer Tucker, "Voyages of Discovery on Oceans of Air: James Glaisher and the Construction of Ballooning as Heroic Scientific Adventure," in Kuklick and Kohler, ed., *Science in the Field*, 144–176.

53. David S. Landes, *Revolution in Time: Clocks and the Making of the Modern World* (Cambridge: Belknap Press of Harvard University Press, 1983); Derek Howse, *Greenwich Time and the Discovery of the Longitude* (New York: Oxford University Press, 1980); Rachel Laudan, *From Mineralogy to Geology: The Foundations of a Science, 1650–1830* (Chicago: University of Chicago Press, 1987); Matthew H. Edney, *Mapping an Empire: The Geographical Construction of British India, 1765–1843* (Chicago : University of Chicago Press, 1990); Robert A. Stafford, *Scientist of Empire: Sir Roderick Murchison, Scientific Exploration and Victorian Imperialism* (Cambridge: Cambridge University Press, 1989). This confidence in surveying was not universal either in Britian or elsewhere, however. Some Canadians, for example, questioned the use of the Geological Survey of Canada.

54. John Cawood, "The Magnetic Crusade: Science and Politics in Early Victorian England," *Isis* 70 (1979): 493–518; Michael S. Reidy, "The Flux and Reflux of Science: The Study of Tides and the Organization of Early Victorian Science" (Ph.D. diss., University of Minnesota, 2000). Within the past two decades, some

historians of science have demonstrated that the field sciences are especially good vehicles for studying the political and economic forces that drive science, the technologies that change the way investigation is done, and the cultural and social traditions that practitioners bring to their fields of study. See Kuklick and Kohler, ed., *Science in the Field.*

55. Lewis Pyenson, *Cultural Imperialism and Exact Sciences: German Expansion Overseas, 1900–1930* (New York: Peter Lang, 1985); Lucille Brockway, *Science and Colonial Expansion: The Role of the British Royal Botanical Gardens* (New York: Academic Press, 1979).

56. Headrick, *Invisible Weapon.*

57. Hydrographic Office, "General Instructions for the Hydrographic Surveyors of the Admiralty" (London: Printed for the HO and sold by J. D. Potter; 1877), HO, Hydrographic Department Publications 65, p. 10. Water is compressible, but not enough to prevent the fall of objects to the sea floor. Untitled newspaper clipping from the *New York Tribune*, July 12, 1845, in IEE, Press Cuttings Relating to Telegraphy, UK0108 SC MSS 021, "Telegraphic Notes, 1839–1857," vol. 1, p. 19; John Murray, *The Ocean: A General Account of the Science of the Sea* (New York: Henry Holt, 1913); William Herdman, *Founders of Oceanography and Their Work: An Introduction to the Science of the Sea* (London: Edward Arnold, 1923), 161.

58. James W. Miller and Ian G. Koblick, *Living and Working in the Sea*, 2nd ed. (Plymouth, VT: Five Corners Publications, 1995), 1–13.

59. A. L. Rice, "The Oceanography of John Ross' Arctic Expedition of 1818: A Reappraisal," *Journal of the Society for the Bibliography of Natural History* 7, no. 3 (1975): 291–319. This recent reexamination of Ross's data indicates that Ross cannot be credited with a 1,000-fathom sounding. He probably reached only 500 or 600 fathoms. Maury, *Physical Geography of the Sea*, 200.

60. Hydrographic Office, "General Instructions," 10.

61. H. M. Denham to John Washington, HMS *Herald*, Feb. 29, 1856, HO, Surveyors Letters, 3d, Denham, 1852.

62. Hydrographic Office (Admiralty), "Atlantic Soundings by H. M. Ships *Herald*, 1852; *Cordelia*, 1868; *Nassau*, 1869; *Valorous*, 1875; *Plover*, 1877; *Alert*, 1878; *Sparrowhawk*, 1886; *Jackal*, 1888," HO, C, Original Document Series, 396. "Sea," in *Encyclopedia Britannica* (1842); reprint of the Royal Institution lecture by John W. Brett, March 20, 1857, in IEE, Press Cuttings Relating to Telegraphy, UK0108 SC MSS 021, "Telegraphic Notes, 1839–1857," vol. 1, tipped in between between pp. 132 and 133.

63. Hydrographic Office, "General Instructions," 10; George E. Belknap, "Deep-Sea Soundings in the North Pacific Ocean Obtained in the U.S. Steamer Tuscarora," U.S. Hydrographic Office, No. 54 (Washington, D.C.: Government Printing Office, 1874), 6–14.

64. Thomson and Murray, ed., *Report of the "'Challenger,"* xxxv; Philip F. Rehbock, "The Early Dredgers: 'Naturalizing' in British Seas, 1830–1850," *Journal of the History of Biology* 12 (1979): 295, 337; A. L. Rice and J. B. Wilson, "The British Association Dredging Committee: A Brief History," in Sears and Merriman, ed., *Oceanography,* 377. In 1842, BAAS dredging committees reported collecting from depths of 50 to 145 fathoms, the deepest accomplished with assistance from

the steam vessel HMS *Lucifer* during its survey of the Scottish coast under Captain F. W. Beechey.

65. Albany Hancock to Joshua Alder, June 26, 1847, and Alder to Rev. Alfred Merle Norman, Sept. 11, 1860, NHM General Library, MSS Ald, vol. 3, 577–578, and vol. 1, 25. All material cited from The Natural History Museum Library collections is used by permission of the Trustees of The Natural History Museum.

66. William Stimpson, "Journal of a Cruise in the US Ship Vincennes to the North Pacific Expedition, Behring Straits, etc.," July 9, 11, 12, 1853, May 2, Dec. 1854, Jan. 9, 24, June 18–20, July 13, Sept. 19, 1855, SIA, NPEE Papers, RU 7253, Box 1; George C. Hyndman to Alder, Belfast, Oct. 15, 1852, NHM General Library, MSS ALD, vol. 4.

67. John G. Jeffreys to Alder, July 28 and July 30, 1861, NHM General Library, MSS ALD, vol. 4, 726–727; J. Gwyn Jeffreys, "The Deep-Sea Dredging Expedition of H.M.S. 'Porcupine': Natural History," *Nature* 1 (1869–1870): 166–168; Jeffreys, "Address to the Biological Section," *BAAS Report*, 1877, pp. 118–119.

68. Alexander Agassiz to Fritz Müller, Oct. 31, 1864, published in Mary P. Winsor, *Starfish, Jellyfish, and the Order of Life: Issues in Nineteenth-Century Science* (New Haven: Yale University Press, 1976), Appendix; E. Percival Wright, "Notes on Sponges," *Quarterly Journal of Microscopical Science*, Jan. 1870, p. 73–81; Thomson, *The Depths of the Sea*, 177.

2. The Undiscovered Country

1. J. H. Parry, *The Discovery of the Sea* (Berkeley: University of California Press, 1981); David S. Landes, *Revolution in Time: Clocks and the Making of the Modern World* (Cambridge: Belknap Press of Harvard University Press, 1983); William J. H. Andrewes, ed., *The Quest for Longitude: The Proceedings of the Longitude Symposium, Harvard University, Cambridge, Massachusetts, November 4–6, 1993* (Cambridge: Collection of Historical Scientific Instruments, Harvard University, 1996); Dava Sobel, *Longitude: The True Story of a Lone Genius Who Solved the Greatest Scientific Problem of His Time* (New York: Walker, 1995).

2. David Mackay, *In the Wake of Cook: Exploration, Science, and Empire, 1780–1801* (New York: St. Martin's Press, 1985); William H. Geotzmann, *Exploration and Empire: The Explorer and the Scientist in the Winning of the American West* (New York: W. W. Norton, 1966); Edward L. Towle, "Science, Commerce, and the Navy on the Seafaring Frontier (1842–1861): The Role of Lieutenant M. F. Maury and the U.S. Naval Hydrographic Office," (Ph.D. diss., University of Rochester, 1966); Goetzmann, *New Lands, New Men: America and the Second Great Age of Discovery* (New York: Penguin Books, 1987); Philip E. Steinberg, *The Social Construction of the Ocean* (Cambridge: Cambridge University Press, 2001), 68–158.

3. Steinberg, *The Social Construction of the Ocean*; Lewis Pyenson, *Cultural Imperialism and Exact Sciences: German Expansion Overseas, 1900–1930* (New York: Peter Lang, 1985); Lucille Brockway, *Science and Colonial Expansion: The Role of the British Royal Botanical Gardens* (New York: Academic Press, 1979); Matthew H. Edney, *Mapping an Empire: The Geographical Construction of British India, 1765–1843* (Chicago: University of Chicago Press, 1990); Michael S.

Reidy, "The Flux and Reflux of Science: The Study of Tides and the Organization of Early Victorian Science" (Ph.D. diss., University of Minnesota, 2000).

4. Matthew Fontaine Maury, *The Physical Geography of the Sea* (New York: Harper & Brothers, 1855), 262–263.

5. Philip L. Richardson, "The Benjamin Franklin and Timothy Folger Charts of the Gulf Stream," in Mary Sears and Daniel Merriman, ed., *Oceanography: The Past* (New York: Springer-Verlag, 1980), 703–717.

6. Richard Ellis, *Men and Whales* (New York: Alfred A. Knopf, 1991); Daniel Francis, *A History of World Whaling* (New York: Viking, 1990); Alexander Starbuck, *History of the American Whale Fishery* (Seacucus, NJ: Castle Books, 1989).

7. Michael T. Bravo, "Science and Discovery in the Admiralty Voyages to the Arctic Regions in Search of a North-West Passage (1815–25)" (Ph.D. diss., Cambridge University, 1992), 73.

8. Sealers ranged as far afield as whalers, and their knowledge of uncharted islands and harbors was superior to that of whalers, who did not normally work on land. But sealers' geographic knowledge held the key to their economic well-being, so they sometimes refused to cooperate with explorers or scientists. Charles Wilkes interviewed whaling and sealing captains while planning the U.S. Exploring Expedition. He found whalers forthcoming, but sealers were unwilling to divulge proprietary information about the location of profitable islands. William Stanton, *The Great United States Exploring Expedition of 1838–1842* (Berkeley: University of California Press, 1975), 17–18; Towle, "Science, Commerce, and the Navy," 449; John Dryden Kazar, "The United States Navy and Scientific Exploration," (Ph.D. diss., University of Massachusetts, 1973), 189, 209–210.

9. Amy Samuels and Peter Tyack, "Flukeprints: A History of Studying Cetacean Societies," in Janet Mann, Richard C. Conner, Peter Tyack, and Hal Whitehead, ed., *Cetacean Societies: Field Studies of Dolphins and Whales* (Chicago: University of Chicago Press, 2000), 11–15.

10. Biographical treatments of Maury include Jaqueline Ambler Caskie, *Life and Letters of Matthew Fontaine Maury* (Richmond, VA: Richmond Press, 1928); Diana F. Maury Corbin, comp., *A Life of Matthew Fontaine Maury* (London: Sampson Low, Marston, Searle, & Rivington, 1888); Charles Lee Lewis, *Matthew Fontaine Maury, the Pathfinder of the Seas* (Annapolis: U.S. Naval Institute, 1927); John Walter Wayland, *The Pathfinder of the Seas: The Life of Matthew Fontaine Maury* (Richmond, VA: Garrett and Massie, 1930); Frances Leigh Williams, *Matthew Fontaine Maury: Scientist of the Sea* (New Brunswick, NJ: Rutgers University Press, 1963); and John Leighly, "M. F. Maury in His Time," *Bulletin de l'Institut Océanographique*, special issue no. 2 (Monaco: Musée Océanographique, 1968), 147–162. Steven J. Dick, *Sky and Ocean Joined: The U.S. Naval Observatory, 1830–2000* (Cambridge: Cambridge University Press, 2003), 60–117.

11. Towle, "Science, Commerce, and the Navy," 141–148; Steven J. Dick, "Centralizing Navigational Technology in America: The U.S. Navy's Depot of Charts and Instruments, 1830–1842," *Technology and Culture* 33 (1992): 469–485. Maury based his sailing directions and wind and current charts, as well as his 1855 book, *Physical Geography of the Sea*, on information gleaned from the log-

books he collected. For an example of his writings based on reports from whalers, see *Explanations and Sailing Directions to Accompany Wind and Current Charts*, 4th ed. (Washington, D.C.: C. Alexander, Printer, 1852), 237–264; hereafter cited as *Sailing Directions*. See D. Graham Burnett, "Matthew Fontaine Maury's 'Sea of Fire': Hydrography, Biogeography, and Providence in the Tropics," in Felix Driver and Luciana Martins, ed., *Tropical Views and Visions* (Chicago: University of Chicago Press, forthcoming).

12. Maury, *Sailing Directions*, 4th ed., 241–245, 255–256, 268.

13. Forbes, *Natural History of the European Seas*, 34, 42–43, 64–65, 76.

14. George C. Wallich, *The North Atlantic Sea-Bed* (London: John van Voorst, 1862), 110; John Murray, "HMS 'Challenger' Diary of Sir John Murray," Jan. 13, 27, 1874, NHM, Mineralogy Library, John Murray Library, Section 1, vol. 1; hereafter cited as "Murray Journal".

15. Joseph Dayman, "Atlantic Soundings, Cyclops, 1857," (London: Published by Order of the Lords Commissioners of the Admiralty; Printed by George E. Eyre and William Spottiswoode, 1858), 4, HO, C. Original Document Series, 391; Frederick J. Evans, "Abstract of Trials for Ocean Soundings in H.M. Ships between the Years 1840 and 1858 and in Ships of the U.S. Navy from 1849 to 1856" (1859), HO, Hydrographic Department Publication 11.

16. Louis Agassiz to John L. LeConte, Dec. 6, [c. 1856], APS, LeConte Papers, B L 493; Maury, *Physical Geography of the Sea*, 146. It is important to note that Maury's scientific work was criticized for not matching the standards of those who worked to professionalize science in the United States, including especially Alexander Dallas Bache, head of the Coast Survey, Joseph Henry, first secretary of the Smithsonian Institution, and Louis Agassiz. Hugh R. Slotten, *Patronage, Practice, and the Culture of American Science: Alexander Dallas Bache and the U.S. Coast Survey* (Cambridge: Cambridge University Press, 1994); Dick, *Sky and Ocean Joined*, 109–117.

17. Mackay, *In the Wake of Cook*; Alan Frost, "Science for Political Purposes: European Explorations of the Pacific, 1764–1806," in Roy MacLeod and Philip F. Rehbock, ed., *Nature in Its Greatest Extent: Western Science in the Pacific* (Honolulu: University of Hawaii Press, 1988), 27–44.

18. E. J. Hobsbawm, *Nations and Nationalism since 1780: Programme, Myth, Reality* (New York: Cambridge University Press, 1990).

19. Geotzmann, *Exploration and Empire*; Goetzmann, *New Lands, New Men*; A. Hunter Dupree, *Science and the Federal Government: A History of Policies and Activities* (Baltimore: Johns Hopkins University Press, 1986), 91–114.

20. Kazar, "The United States Navy and Scientific Exploration," vi–ix; Towle, "Science, Commerce, and the Navy," 15–75; Harold L. Burstyn, "Seafaring and the Emergence of American Science," in Benjamin W. Labaree, ed., *The Atlantic World of Robert G. Albion* (Middletown, CT: Wesleyan University Press, 1975), 101; Stanton, *The Great United States Exploring Expedition*; Herman Viola and Carolyn Margolis, *Magnificent Voyagers: The U.S. Exploring Expedition, 1838–1842* (Washington, D.C.: Smithsonian Institution Press, 1985); Goetzmann, *New Lands, New Men*, 270–297.

21. Rebecca Ullrich, "British Scientists at Sea: Naturalists on Naval Vessels, 1830–1870," paper delivered at the History of Science Society Annual Meeting, Santa

Fe, NM, Nov. 1993; Alfred Friendly, *Beaufort of the Admiralty: The Life of Sir Francis Beaufort, 1774–1857* (London: Hutchinson, 1977); G. S. Ritchie, *The Admiralty Chart: British Naval Hydrography in the Nineteenth Century* (London: Hollis & Carter, 1967).

22. Benjamin Franklin, *Journal of a Voyage* (1726); see Sept. 2, 9, 28, 29, 30 entries; Charles Darwin, *The Voyage of the "Beagle"* (New York: Penguin Books, 1988), Dec. 6, 1833, pp. 135–141. Civilian marine naturalists who went to sea on naval vessels before the 1860s include Edward Forbes, William Stimpson, Louis Agassiz, and George Wallich. Of these, only Stimpson traveled with an official exploring expedition; the rest sailed on surveying vessels.

23. John Murray and John Hjort, *The Depths of the Ocean* (London: Macmillan, 1912), 2–5.

24. Credible efforts were made in the eighteenth century to sound Arctic waters; Margaret Deacon, *Scientists and the Sea, 1650–1900: A Study of Marine Science*, 2nd ed. (Brookfield, VT: Ashgate Publishing Co., 1997), 228–229, 281–283, 309. For more information about Ross's expedition, see Ritchie, *The Admiralty Chart;* A. L. Rice, "The Oceanography of John Ross' Arctic Expedition of 1818: A Reappraisal," *Journal of the Society for the Bibliography of Natural History* 7, no. 3b(1975): 291–319.

25. Louis Agassiz to Franklin Haven, Esq., July 15, 1871, APS, Miscellaneous MSS. Collection.

26. Towle, "Science, Commerce, and the Navy," 15–75.

27. The quote is from Smithsonian Institution, *Annual Report,* 1854, p. 89. Historians have characterized the expedition as a footnote to Matthew Perry's opening of Japan to Western commerce, but few have written specifically about it. One exception is Alan Burnett Cole, *Yankee Surveyors in the Shogun's Seas: Record of the United States Surveying Expedition to the North Pacific Ocean, 1853–1856* (Princeton: Princeton University Press, 1947). From the point of view of history of science, the best treatment of the expedition is a chapter in John D. Kazar's dissertation about U.S. naval expeditions, although a manuscript biography of Stimpson by Ronald Vasile includes several chapters about the expedition. Towle wrote about the Arctic portion of the expedition, and Goetzmann also covers it in *New Lands, New Men.* Kazar, "The United States Navy and Scientific Exploration, 1837–1860"; Ronald Vasile, "A Fine View from the Shore: The Life of William Stimpson," unpublished manuscript; Towle, "Science, Commerce, and the Navy," 423–474; Goetzmann, *New Lands, New Men,* 349–357. The expedition departed on June 11, 1853. It did not return to the east coast of the United States until July 11, 1856. Almost no mention of collecting was made after September 1855, when the expedition left Japanese waters to cross the North Pacific to the Aleutian Islands, from which point it sailed to San Francisco. Stimpson kept a daily journal of the expedition; see "Journal of a Cruise in the US Ship Vincennes to the North Pacific Expedition, Behring Straits, Etc.," SIA, NPEE, RU 7253, Box 1; hereafter cited as "Stimpson Journal".

28. William Stimpson to J. H. Dana, Washington, Dec. 1, 1856, published in "Miscellaneous Scientific Intelligence," in *American Journal of Science and Arts,* 2nd ser., 23 (1857): 136–138.

29. Towle, "Science, Commerce, and the Navy," 423–474.

30. Kazar, "The United States Navy and Scientific Exploration," 189.
31. Edward Lurie, *Louis Agassiz: A Life in Science* (Baltimore: Johns Hopkins University Press, 1988); Slotten, *Patronage, Practice, and the Culture of American Science*, 134–136; "Stimpson Journal," June 13, July 1, 3, 8, 10, 12, 22, 29, Aug. 7, Nov. 9, Dec. 22, 1853, Feb. 14, May 3, Nov.–Dec. 1854, and April 12, 1855.
32. Just a few years earlier, in 1841, Forbes dredged in the Mediterranean while on the hydrographic cruise of HMS *Beacon*. Given dredging practices at the time in Britian, it is likely that Forbes dredged from boats rather than the ship itself. "Stimpson Journal," April 12, Nov. 28, 1854, and May 12, 1855.
33. The quote is from "Stimpson Journal," Jan. 9, 1855. Although he brought with him an arsenal of collecting implements, Stimpson observed and borrowed the methods and tools of native fishermen; see April 1, May 4, 1854, and June 12, 1855. Some of the dredging events described in Stimpson's journal do not record the depth, but describe the ship or boat as being farther out than usual, and thus probably also in deeper water than his more typical collecting grounds; see July 9, 11, 1853, May 3, Dec. 1854, Jan. 24, June 18–20, July 13, Sept. 19, 1855.
34. George M. Brooke, Jr., *John M. Brooke: Naval Scientist and Educator* (Charlottesville: University Press of Virginia, 1980), 99. Towle says that Maury engineered this transfer. Towle, "Science, Commerce, and the Navy," 38–52, 442–447.
35. Brooke, *John M. Brooke*, 75–84.
36. "Stimpson Journal," July 24, Nov. 15, 1853, Oct. 13, Dec. 25, 1854, April 18, July 19, 1855; Brooke, *John M. Brooke*, 93–94; George M. Brooke, Jr., *John M. Brooke's Pacific Cruise and Japanese Adventure, 1858–1860* (Honolulu: University of Hawaii Press, 1986), Jan. 31 1854; Stimpson to Commander John Rodgers, Washington, Oct. 3, 1860, SIA, NPEE, RU 7253, Box 1, folder 15.
37. Charles Wyville Thomson, *The Depths of the Sea: An Account of the General Results of the Dredging Cruises of HMSS "Porcupine" and "Lightning" during the Summers of 1868, 1896, and 1870* (London: Macmillan., 1873), 224. In claiming the *Porcupine*'s 1869 sounding as the first truly accurate one, Thomson dismissed, without comment, a number of other claimants discussed in Chapter 3.
38. Kazar, "The United States Navy and Scientific Exploration," 221–222, 224. The only cruise narrative published was A. W. Habersham, *The North Pacific Surveying and Exploring Expedition, or My Last Cruise* (Philadelphia: Lippincott, 1858). Others have written brief accounts of the expedition, including Goetzmann, *New Lands, New Men*, 349–356, and Towle, "Science, Commerce, and the Navy," 423–467. See William A. Deiss and Raymond B. Manning, "The Fate of the Invertebrate Collections of the North Pacific Exploring Expedition, 1853–1856," in *History in the Service of Systematics* (London: Society for the Bibliography of Natural History, 1981), 79–85.
39. Kazar, "The United States Navy and Scientific Exploration," 222.
40. Brooke, *John M. Brooke*, 151–156.
41. William Stimpson to Spencer Baird, March 11, 1860, March 20, 1860, July 12, 1861, Aug. 21, 1863, and May 23, 1870, all in SIA, SFB, RU 7002, Box 33. Stimpson to William Healey Dall, Academy of Science, Chicago, June 15, 1871, SIA, WHD, RU 7073, Box 3.

42. Stimpson to Baird, July 12, 1861, SIA, RU 7002, SFB, Box 33; Edward F. Rivinus and E. M. Youssef, *Spencer F. Baird of the Smithsonian* (Washington, D.C.: Smithsonian Institution Press, 1992), 81–97, 141–151.

43. Ullrich, "British Scientists at Sea"; Kazar, "The United States Navy and Scientific Exploration," 210; Eric L. Mills, "Edward Forbes, John Gwyn Jeffreys, and British Dredging before the *Challenger* Expedition," *Journal of the Society for the Bibliography of Natural History* 8, no. 4 (1978): 507–536; Philip F. Rehbock, "Edward Forbes: An Annotated List of Published and Unpublished Writings," *Journal of the Society for the Bibliography of Natural History* 9, no. 2 (1979): 71–218.

44. Agassiz commented to Baird in 1870 that in the United States it still helped to be a physician to be invited along as an expedition naturalist. Agassiz to Baird, Oct. 13, 1870, in Elmer Charles Herber, ed., *Correspondence between Spencer Fullerton Baird and Louis Agassiz: Two Pioneer American Naturalists* (Washington, D.C.: Smithsonian Institution Press, 1963); Ullrich, "British Scientists at Sea."

45. The fact that civilian scientists were included on naval expeditions did not indicate, however, that they were welcomed by all naval personnel. Kazar, "The United States Navy and Scientific Exploration," 270–273.

46. The quote is from George C. Wallich, "Daily Diary of Voyage of 'Bulldog'," July 2, 1860, NHM General Library, MSS WAL; hereafter cited as "Wallich Journal."; Wallich to Joseph Hooker, Nov. 6, 1874, at KEW, Letters to J. D. Hooker, vol. 21, 41–43.

47. Francis Darwin, *More Letters of Charles Darwin* 1(1903), 70. Wallich to Hooker, Kensington, June 1, 1860, KEW, English Letters (1857–1900), vol. 104: 348.

48. The quote is from George Wallich, "Lett's Diary for 1860," entries from May 15–June 1, 1860, NHM General Library, MSS WAL; hereafter cited as "Wallich Diary, 1860"; "Wallich Journal," July 2, 1860; Robert A. Stafford, *Scientist of Empire: Sir Roderick Murchison, Scientific Exploration, and Victorian Imperialism* (Cambridge: Cambridge University Press, 1989). Of the many excellent biographies of Darwin, one that particularly emphasizes the importance of his *Beagle* voyage is Janet Browne, *Charles Darwin: Voyaging* (New York: Alfred A. Knopf, 1995).

49. A. L. Rice, Harold L. Burstyn, and A. G. E. Jones, "G. C. Wallich, MD: Megalomaniac or Mis-Used Oceanographic Genius?" *Journal of the Society for the Bibliography of Natural History* 7, no. 4 (1976): 423–450. This article says that Wallich had three children at the time, but the list of children with their birthdays in note 5 indicates that there could have only been two then.

50. The quotes are from "Wallich Journal," July 2 and 12, 1860, respectively; see also Sept. 9, 1860. Professor Bell could be Joseph Bell (1817–1863), who was a physician and professor of botany at Anderson College, Glasgow, from 1847 on. He received his M.D. degree in Edinburgh in 1832, four years before Wallich earned his M.D. there. Rice, Burstyn, and Jones, "G. C. Wallich," 424. The slides are held in the Paleontology Department, NHM, and are still used by researchers.

51. George Wallich, "Private Diary for 1861," Jan. 25, 1861, NHM General Library, MSS WAL; hereafter cited as "Wallich Diary, 1861".

52. The quote is from Wallich to Hooker, March 12, 1866, KEW, Letters to J. D.

Hooker, vol. 21, 26; Sir Roderick Murchison to Wallich, May 18, 1865, NHM General Library, MSS WAL, Box 1, "Murchison"; Wallich to Hooker, Kensington, Feb. 21, 1867, Feb. 22, 1867, and Feb. 25, 1867, KEW, Letters to J. D. Hooker, vol. 21: 27–28, 29, 30. Unfortunately, Wallich failed financially at this photographic venture, after investing heavily in a picture book of well-known scientists of the day. Indeed, he went bankrupt several times during his life, although he seems to have patented some improved microscopes and other scientific equipment. See Rice, Burstyn, and Jones"G. C. Wallich," 423–450.

53. On his attempts to sanction Carpenter and Thomson, see the series of letters between Wallich and Hooker beginning in December 1877, at RSA, MC 11, 126–150. For the sounding-device dispute, see "Wallich Diary, 1861," January to June. Helen M. Rozwadowski, "Technology and Ocean-Scape: Defining the Deep Sea in the Mid Nineteenth Century," *History and Technology* 17 (2001): 217–247.

54. William B. Carpenter to Sir Edward Sabine, Nov. 3, 1869, RSA, Sa. 301.

55. Richard Henry Dana, *Two Years before the Mast* (Cleveland: World Publishing Co., 1946), 312–314.

56. John Cawood, "The Magnetic Crusade: Science and Politics in Early Victorian England," *Isis* 70 (1979): 493–518; Michael S. Reidy, "The Spaces in Between: British Science and Imperial Oceans," paper delivered at the Environmental History and the Oceans Conference sponsored by the German Historical Intitute, held at the Carlsberg Academy, Copenhagen, Denmark, June 2–5, 2004.

57. Forbes, *Natural History of the European Seas,* 10–11.

58. C. G. Ehrenberg to Matthew F. Maury, Berlin, Oct. 1857, in Maury, *Sailing Directions,* 8th ed. (Washington: William A. Harris, Printer, 1858–59), 175; William Stimpson to Commander Ringgold, Washington, D.C., Dec. 15, 1852, SIA, NPEE, RU 7253, Box 1, folder 7.

59. Maury, *Sailing Directions,* 5th ed. (Washington, D.C.: C. Alexander, Printer, 1853), 212, 219–220; Maury, *Sailing Directions,* 4th ed., 302; J. C. Dobbin, "Annual Report of the Secretary of the Navy," in Diana F. Maury Corbin, comp., *A Life of Matthew Fontaine Maury* (London: Sampson Low, Marston, Searle, & Rivington, 1888), 101–103.

60. Ritchie, *The Admiralty Chart,* 3; John A. Edgell, *Sea Survey: Britain's Contribution to Hydrography* (London: HMSO, 1965). The Maury quote is from Williams, *Matthew Fontaine Maury,* 148.

61. Sir W. Armstrong, "Address to the Mechanial Science Section," in "Transactions of the Sections" of *BAAS Reports,* 1865, p. 164.

3. Soundings

1. George C. Wallich, "Daily Diary of Voyage of 'Bulldog,," Aug. 22, 1860, NHM General Library, MSS WAL; hereafter cited as "Wallich Journal."

2. Horace Beck, *Folklore of the Sea* (Middletown, CT: Wesleyan University Press, 1973), 108–109.

3. Instructions written for British hydrographers emphasized the importance of reliable soundings to fogbound navigators. The first set of instructions was drawn up by Beaufort, with the first printing in 1850. Revisions were published in 1858,

1862, and again in 1877. See Hydrographic Office, "General Instructions for the Hydrographic Surveyors of the Admiralty" (1877), 8, 10, 31, HO, Hydrographic Department Publication 65.

4. Dirk J. Struik, *Yankee Science in the Making: Science and Engineering in New England from Colonial Times to the Civil War* (New York: Dover Publications, 1991), 93–134.

5. Harold Burstyn, "Seafaring and the Emergence of American Science," in Benjamin W. Labaree, ed., *The Atlantic World of Robert G. Albion* (Middletown, CT: Wesleyan University Press, 1975), 76–109.

6. Hugh R. Slotten has written an excellent history of the Survey under Bache, *Patronage, Practice, and the Culture of American Science: Alexander Dallas Bache and the U.S. Coast Survey* (Cambridge: Cambridge University Press, 1994). See also A. Hunter Dupree, *Science and the Federal Government: A History of Policies and Activities* (Baltimore: Johns Hopkins University Press, 1986).

7. Steven J. Dick, "Centralizing Navigational Technology in America: The U.S. Navy's Depot of Charts and Instruments, 1830–1842," *Technology and Culture* 33 (1992): 469–485; Steven J. Dick, *Sky and Ocean Joined: The U.S. Naval Observatory, 1830–2000* (Cambridge: Cambridge University Press, 2003), 60–117; Edward L. Towle, "Science, Commerce, and the Navy on the Seafaring Frontier (1842–1861): The Role of Lieutenant M. F. Maury and the U.S. Naval Hydrographic Office" (Ph.D. diss., The University of Rochester, 1966), 118–125; Thomas G. Manning, *U.S. Coast Survey vs. Naval Hydrographic Office: A 19th-Century Rivalry in Science and Politics* (Tuscaloosa: University of Alabama Press, 1988). For a refreshing reassessment of Maury's work, see D. Graham Burnett, "Matthew Fontaine Maury's 'Sea of Fire': Hydrography, Biogeography, and Providence in the Tropics," in Felix Driver and Luciana Martins, ed., *Tropical Views and Visions* (Chicago: University of Chicago Press, forthcoming).

8. Frances Leigh Williams, *Matthew Fontaine Maury: Scientist of the Sea* (New Brunswick, NJ: Rutgers University Press, 1963), 140.

9. Matthew Fontaine Maury, *Explanations and Sailing Directions to Accompany Wind and Current Charts*, 4th ed. (Washington: C. Alexander, Printer, 1852), 128; hereafter cited as *Sailing Directions, 4th ed.*; S. P. Lee to Commander Morris, Dec. 20, 1851, in Maury, *Sailing Directions*, 4th ed., 175–176.

10. "Telegraph Supplement," *Harper's Weekly*, Sept. 4, 1858, p. 10.

11. Because they thought water was compressible, they expected seawater at great depths to have increased density (to be more "dense"). In fact, seawater is compressible, although its compressibility is only about 4 percent below 4,000 meters. H. U. Sverdrup, Martin W. Johnson, and Richard H. Fleming, *The Oceans: Their Physics, Chemistry, and General Biology* (Englewood Cliffs, NJ: Prentice-Hall, 1942), 68. Charles Wyville Thomson refers to this belief in *The Depths of the Sea: An Account of the General Results of the Dredging Cruises of HMSS "Porcupine" and "Lightning" during the Summers of 1868, 1896, and 1870* (London: Macmillan, 1873), 31–32.

12. Untitled newspaper clipping from the *New York Tribune*, July 12, 1845, in IEE, Press Cuttings Relating to Telegraphy, UK0108 SC MSS 021, "Telegraphic Notes, 1839–1857, vol. 1, 19; Francis Higginson, *The Ocean: Its Unfathomable Depths and Natural Phenomena: Comprising Authentic Narratives and Strange*

Reminisences of Enterprise, Delusion, and Delinquency: with the Voyage and Discoveries of HMS "Cyclops" (London: Edward Stanford, 1857), 6–13. Captain Robert Fitzroy, who commanded the HMS *Beagle* voyage that Darwin sailed on, maintained his belief that depths could not be sounded even after experiments proved that water density did not increase appreciably with depth. See Theodore F. Jewell, "Deep Sea Sounding," *The Record of the United States Naval Institute* [Proceedings of the U.S. Naval Institute] 4, no. 3(1877): 37.

13. The textbook was Maury's *New and Practical Treatise on Navigation* (Philadelphia: Key and Biddle, 1836). The wind and current investigations were summarized and also explained in detail in sequential editions of *Sailing Directions* published from 1851 to at least 1858. Although there is some overlap of material between editions, most of them contain distinct information. The Act of Congress, March 3, 1849, authorized wind and current studies, but Maury interpreted it broadly by pursuing studies of whale distribution and deep-sea bathymetry. Maury, *Sailing Directions*, 4th ed., 125. For an account of the *Taney* cruise, see Towle, "Science, Commerce, and the Navy," 230–238, 254–273.

14. Maury, *Sailing Directions*, 4th ed., first quote on p. 54, second on p. 162.

15. Matthew Fontaine Maury, *Lieut. Maury's Investigations of the Winds and Currents of the Sea* (Washington, D.C.: C. Alexander, Printer, 1851), 30. This conclusion appeared in later editions, such as *Sailing Directions*, 4th ed., 130–132. Walsh's report was published as J. C. Walsh to M. F. Maury, Aug. 15, 1850, Appendix B, in M. F. Maury, *U.S. National Observatory Astronomical Observations*, vol. 2 (Washington, D.C., 1851), 89–137.

16. George E. Belknap wrote that the *Taney*'s was not the first wire sounding, in "Something about Deep-Sea Sounding," *The United Service* 2 (April 1879): 162–163. Although wire was not successfully used in the 1840s, it was in the late 1870s; Maury, *Sailing Directions*, 4th ed., 131–132. The *Taney* carried so much continuous wire on one reel not because organizers of the cruise expected such spectacular depths, but because their sounding method consisted of cutting loose the line and sinker after the sounding. Since hydrdrographers did not recover the line, and since wire was easier to store and deploy over the side when wound on reels rather than stored as individual lengths, all the wire was carried on one reel or a few large reels.

17. The "dead" quote is Walsh's, as quoted in Towle, "Science, Commerce, and the Navy," 258 (italics in the original manuscript log of voyage). The "desirable" quote is from Maury, *Lieut. Maury's Investigations*, 30. Other quotes are from Maury, *Sailing Directions*, 4th ed., 131.

18. Circular, Bureau of Ordnance and Hydrography, May 31, 1850, printed in Maury, *Sailing Directions*, 3rd ed. (Washington, D.C.: C. Alexander, Printer, 1851), 70–71.

19. Maury, *Sailing Directions*, 6th ed. (Philadelphia: E. C. and J. Biddle, 1854), 231; Maury, *Sailing Directions*, 8th ed. (Washington, D.C.: William A. Harris, Printer, 1858), 125.

20. Maury, *Sailing Directions*, 4th ed., 140–170.

21. Maury, *Sailing Directions*, 4th ed., 130–140, 184; Maury, *Sailing Directions*, 5th ed. (Washington, D.C.: C. Alexander, Printer, 1855), 207–210, 228–229; Maury, *Sailing Directions*, 8th ed., 142.

22. Samuel Phillips Lee, *Report and Charts of the Cruise of the US Brig "Dolphin"* (Washington, D.C.: Beverley Tucker, Printer to the Senate, 1854); Dudley Taylor Cornish and Virginia Jeans Laas, *Lincoln's Lee: The Life of Samuel Phillips Lee, United States Navy, 1812–1897* (Lawrence: University Press of Kansas, 1986). Towle describes *Dolphin*'s voyage in "Science, Commerce, and the Navy," 276–281.

23. Lee, *Cruise of the "Dolphin,"* 9, 19–22, 34, 113, 180.

24. Maury, *Sailing Directions,* 4th ed., 138–140.

25. The quote is from S. P. Lee to Morris, in Maury, *Sailing Directions,* 4th ed., 175; Lee, *Cruise of the "Dolphin,"* 330.

26. Maury, *Sailing Directions,* 5th ed., 191; Maury, *Sailing Directions,* 7th ed. (Washington, D.C.: William A. Harris, Printer, 1855), 134–137.

27. During the next decade, hydrographers less often blamed this effect on undercurrents, because they realized that the weight of thousands of fathoms of sounding line would pull additional line off the reel even without the action of currents.

28. Maury, *Sailing Directions,* 5th ed., 191; Maury, *Sailing Directions,* 7th ed., 130–133.

29. The quote is from Maury, *Sailing Directions,* 4th ed., 138; Maury, *Sailing Directions,* 7th ed., 123.

30. The quote is from Maury to Cyrus W. Field, Nov. 3, 1853, NA, Record Group 78, Entry 1 [Letterbooks 1842–1892], vol. 10, 23–34; Maury, *Lieut. Maury's Investigation,* 52. *Physical Geography of the Sea* was the title of his popular book, whose content was derived from his many editions of *Sailing Directions*. Susan Faye Cannon, *Science in Culture: The Early Victorian Period* (New York: Dawson and Science History Publications, 1978).

31. Maury, *Sailing Directions,* 6th ed., Plate XV. Maury noted that the land surface profile was taken from "Dr. Drake's work on 'The Principle Diseases in the Interior Valley of North America'"; Maury, *Sailing Directions,* 5th ed., 240, Plate XV. There were eighteenth-century precedents for using submarine contours in relatively shallow waters; see Jacqueline Carpine-Lancre, "The Origin and Early History of *la carte générale bathymétrique des oceans*," in Jacqueline Carpine-Lancre, Robert Fisher, Brian Harper, Peter Hunter, Meirion Jones, Adam Kerr, Anthony Laughton, Steve Ritchie, Desmond Scott, and Maya Whitmarsh, ed., *The History of GEBCO, 1903–2003: The 100-Year Story of the General Bathymetric Chart of the Oceans* (Lemmer, The Netherlands: Goematics Information and Trading Centre, 2003), 16.

32. Maury, *Sailing Directions,* 5th ed., 226.

33. The quote is from Maury, *Sailing Directions,* 7th ed., 154; Maury, *Sailing Directions,* 5th ed., 240; Maury, *Sailing Directions,* 6th ed., Plate XV, 298.

34. Maury, *Sailing Directions,* 6th ed., 216–217.

35. Maury to Morse, Feb. 23, 1854, and Maury to Field, Nov. 3, 1853, NA, Record Group 78, Entry 1 [Letterbooks 1842–1892], vol. 10, 218–220, 23–34.

36. Maury to Hon J. C. Dobbin, Secretary of the Navy, Nov. 8, 1856, NA, Record Group 78, Entry 1 (Letterbooks 1842–1892), vol. 14, 36–50; Maury, *Sailing Directions,* 8th ed., 157–158.

37. Maury to Dobbin, Feb. 22, 1854, in Jaquelim Ambler Caskie, *Life and Letters of*

Matthew Fontaine Maury (Richmond: Richmond Press, 1928), 110–112 (italics in the original). All available evidence supports the conclusion that Maury had already composed a letter to the Navy Secretary asserting the feasibility of a cable across the Atlantic before he received Field's letter. Maury and Samuel Morse, with whom Maury and Field both corresponded about an Atlantic telegraph at this time, were friends, so it is likely that Morse first suggested to Maury the idea of oceanic telegraphy, which he predicted in the 1840s. Morse became a consultant of Field's New York, Newfoundland, and London Telegraph Company. See Henry M. Field, *The Story of the Atlantic Cable* (New York: Charles Scribner's Sons, 1892), and Maury to Cyrus W. Field, Feb. 24, 1854, NA, Record Group 78, Entry 1 [Letterbooks 1842–1892], vol. 10, 222.

38. Maury to Dobbin, Feb. 22, 1854, in Caskie, *Life and Letters of Matthew Fontaine Maury*, 110–112.

39. George M. Brooke, Jr., *John M. Brooke: Naval Scientist and Educator* (Charlottesville: University Press of Virginia, 1980), 38–52.

40. Brooke, *John M. Brooke*, 54–58; Maury, *Sailing Directions*, 8th ed., 125.

41. Maury, *Sailing Directions*, 5th ed., 238–239 (quotes are from here), Plate XIV; Maury, *Sailing Directions*, 7th ed., plate XIV.

42. Maury, *Sailing Directions*, 8th ed., 157–158.

43. The quote is from Maury, *Sailing Directions*, 8th ed., 157–158; Maury to Prof. J. W. Bailey, Nov. 25, 1853, and Maury to Prof. Ehrenberg, Nov. 25, 1853, NA, Record Group 78, Entry 1 [Letterbooks 1842–1892], vol. 10, 66, 67; Maury, *Sailing Directions*, 7th ed., 155; Stanley Coulter, "Jacob Whitman Bailey," *Botanical Gazette* 13 (1888): 118–124.

44. Maury, *Physical Geography of the Sea*, 209; Maury, *Sailing Directions*, 7th ed., 154–155. Here, Maury quoted Benjamin Franklin on his observations of the first balloon ascent in pre-Revolutionary France.

45. R. J. Mann, *The Atlantic Telegraph: A History of Preliminary Experimental Proceedings, and a Descriptive Account of the Present State and Prospects of the Undertaking* (London: Jarrold and Sons, July 1857), 32–33.

46. Maury to Prof. S. F. B. Morse, Feb. 23, 1854, and Maury to Cyrus W. Field, Feb. 24, 1854, NA, Record Group 78, Entry 1 [Letterbooks 1842–1892], vol. 10, 218–220, 222; Williams, *Matthew Fontaine Maury*, 225–257. For examples of work on the electrical science and engineering of cables, see Bruce J. Hunt, "Michael Faraday, Cable Telegraphy, and the Rise of Field Theory," *History of Technology* 13 (1991): 1–19, and Iwan Rhys Morus, "Currents from the Underworld: Electricity and the Technology of Display in Early Victorian England," *Isis* 84, no. 1(1993): 50–69.

47. Maury to Isaac Toucey, Secretary of the Navy, Sept. 4, 1858, NA, Record Group 78, Entry 1 [Letterbooks 1842–1892], vol. 16, 21–33.

48. Clipping from the *New York Herald*, April 20, 1857, in IEE, Press Cuttings Relating to Telegraphy, UK0108 SC MSS 021, Telegraphic Notes, 1839–1857," vol. 1; and "The Recent Soundings for the Atlantic Telegraph," *The Illustrated London News* [Fall 1857], in IEE, Press Cuttings Relating to Telegraphy, UK0108 SC MSS 021, "Telegraphic Notes, 1857–59," vol. 3, 32.

49. The quote is from "The Atlantic Telegraph Co.," *The Times*, Nov. 14, 1856, in IEE, Press Cuttings Relating to Telegraphy, UK0108 SC MSS 021, "Telegraphic

Notes, 1839–1857," vol. 1, 115; Cyrus W. Field, "The Atlantic Telegraph," *Proceedings of the Royal Geographic Society* (London), Dec. 1856/Jan. 1857, pp. 216–217.

50. Quotes are from "The Atlantic Telegraph: A History of Preliminary Experimental Proceedings, and a Descriptive Account of the Present State and Prospects of the Undertaking," published by order of the Directors of the Company (London: Jarrold and Sons, July, 1857), frontispiece, Plate II, and pp. 27–32, C&WA. Cable-laying efforts inspired dozens of narratives, ranging from blow-by-blow stories about the voyages during which cables were laid to "histories" that were, in fact, pleas to bestow on engineers, electricians, or entrepreneurs the credit they deserved and were not, for whatever reason, getting. Most of all, these publications were thinly veiled promotional pieces designed to attract investors and boost public confidence.

51. The Submarine Telegraph, *The Story of My Life* (London: C. West, 1859), 68.

52. The first quotes are from Maury to Dobbin, Feb. 22, 1854, in Corbin, *A Life of Matthew Fontaine Maury,* 99–101 (italics in the original); Maury, *The Physical Geography of the Sea,* 212.

53. Corbin, *A Life of Matthew Fontaine Maury,* 104.

54. Great Britain, Submarine Telegraph Company, *Report of the Joint Committee Appointed by the Lords of the Committee of Privy Council for Trade and the Atlantic Telegraph Company to Inquire into the Construction of Submarine Telegraph; Together with the Minutes of Evidence and Appendix* (London: George Eyre and William Spottiswoode, for HMSO, 1861), xxix–xxxv, 8–9, 17, 24, 47, 53, 181–185, 188, 193, 208–215, 221–225, 245–246, 253.

55. "Miscellaneous Scientific Intelligence," *The American Academy of Science and Arts,* 2nd ser., 26, no. 77 (1858): 285.

56. "Wallich Journal," July 6, July 7, Sept 9, Aug. 19, 1860; "The North Atlantic Telegraph via the Faröe Isles, Iceland, and Greenland: Preliminary Reports of the Surveying Expeditions of 1860" (London: Edward Stanford, 1861), 2, SI, DPH, Judson Acc. 32290, Cat. 7248.

57. Atlantic Telegraph Company, "Report of the Proceedings of the Extraordinary General Meeting at the London Tavern on Friday, Dec. 12, 1862" (London, 1863), 9–10, SI, DPH, Judson Acc. 32290, Cat. 7248; Cyrus W. Field, "Prospects of the Atlantic Telegraph: A Paper Read before the American Geographical and Statistical Society, May 1, 1862," SI, DPH, Judson Acc. 32290, Cat. 7248; Arturo de Marcoartu, Universal Telegraphic Enterprise, "Telegraphic Submarine Lines between Europe and America, and the Atlantic and Pacific" (New York: S. Hallet, Book & Job Printer, 1863), 8, SI, DPH, Judson Acc. 32290, Cat. 7248.

58. Lee, *Report and Charts of the Cruise of the US Brig "Dolphin."* For example, Captain W. J. Pullen referred to the *Dolphin*'s cruise in "Atlantic Soundings and Temperatures (Serial) by HMS Cyclops, 1857–8," 9, HO, C., Original Document Series, 392; Helen M. Rozwadowski, "Technology and Ocean-Scape: Defining the Deep Sea in the Mid Nineteenth Century," *History and Technology* 17 (2001): 217–247.

59. G. S. Ritchie, *The Admiralty Chart: British Naval Hydrography in the Nineteenth Century* (London: Hollis & Carter, 1967); John F. W. Herschel, ed., *A*

Manual of Scientific Inquiry; Prepared for the Use of Officers in Her Majesty's Navy; and Travelers in General (London: John Murray, 1849).

4. A Sea Breeze

1. Mary P. Winsor, *Reading the Shape of Nature: Comparative Zoology at the Agassiz Museum* (Chicago: University of Chicago Press, 1991); David Allen, *The Naturalist in Britain: A Social History* (Princeton: Princeton University Press, 1994), 122, 129–310; Anne L. Larsen, "Not Since Noah: The English Scientific Zoologists and the Craft of Collecting, 1800–1840," (Ph.D. diss., Princeton University, 1993), 129–195.
2. L. F. Marsigli, *Histoire physique de la mer* [The Natural History of the Sea], trans. with Introduction and notes by Anita McConnell, with the assistance of Jacqueline Carpine-Lancre, Piero Todesco, and Giorgio Tabarroni, a second Introduction by Giorgio Dragoni (Bologna: Museo di Fisica dell'Università di Bologna, 1999). A seventeenth-century Norwegian dredger was Peder Ascanius; Liv Bliksrud, Geir Hestmark, and Tarald Rasmussen, *Vitenskapens utfordringer, 1850–1920* [The Challenges of Science, 1850–1920], Norsk idéhistorie [Norwegian History of Ideas], vol. 4 (Oslo: Aschoug, 2002), 23–69. Eric L. Mills, "Edward Forbes, John Gwyn Jeffreys, and British Dredging before the *Challenger* Expedition," *Journal of the Society for the Bibliography of Natural History* 8, no. 4 (1978): 507–536; Philip F. Rehbock, "The Early Dredgers: Naturalizing in British Seas, 1830–1850," *Journal for the History of Biology* 12, no. 2 (1979): 293–368.
3. Charles Wyville Thomson and John Murray, ed., *Report of the Scientific Results of the Exploring Voyage of the HMS "Challenger," 1873–76: Narrative of the Cruise,* vol. 1 (London: HMSO, 1885), xxxv. See also Rehbock, "Early Dredgers," 295, and Allen, *Naturalist in Britain,* 142–157.
4. Rehbock, "Early Dredgers," 297–306, 355.
5. Edward Forbes, "On the Associations of Mollusca and the British Coasts, Considered with Reference to Pleistocene Geology," *Edinburgh Academic Annual* (1840), pp. 175–183. Edward Forbes's "Dredging Song" is published in full (four stanzas) in Rehbock, "Early Dredgers," 326–327. The "To afford" quote is in E. Perceval Wright, "Scientific Opinion: Notes on Deep Sea Dredging," *Annals and Magazine of Natural History,* Dec. 16, 1868, clipping in NHM General Library, MSS WAL. Christian Ehrenberg to [unknown], Dec. 17, 1860, NHM General Library, MSS WAL.
6. Quoted in Rehbock, "Early Dredgers," 306.
7. George Wallich to Sir Archibald Geike, March 30, 1878, EUL, Gen. 526; Charles Wyville Thomson, *The Depths of the Sea: An Account of the General Results of the Dredging Cruises of HMSS "Porcupine" and "Lightning" during the Summers of 1868, 1896, and 1870* (London: Macmillan, 1873), 241; Robert Patterson, "Memoir of the Late Robert Ball, LL.D., M.R.I.A.," *The Natural History Review* 5 (1858): 1–34; *BAAS Report* (1849), p. 72; Rehbock, "Early Dredgers," 295.
8. Edward Forbes, "On the Dredge," appendix to Alfred Tulk and Arthur Henfry, *Anatomical Manipulation* (London: John van Voorst, 1844), 405–407; Robert Ball, "Notice of a New Dredge for Natural-History Purposes," *BAAS Report*

(1849), p. 72; Louis F. de Pourtales, "Directions for Dredging," *Bulletin of the Museum of Comparative Zoology at Harvard College* 2 (1870–71): 451–454. Ball's dredges came in two sizes: the larger one was 3 feet × 6 inches while the smaller was 18 inches × 6 inches.

9. Ronald Vasile, "A Fine View from the Shore: The Life of William Stimpson" unpublished manuscript; Christopher Hamlin, "Robert Warington and the Moral Economy of the Aquarium," *Journal of the History of Biology* 19 (1986): 131–153. British dredgers also kept animals alive in seawater tanks. See Philip F. Rehbock, "The Victorian Aquarium in Ecological and Social Perspective," in Mary Sears and Daniel Merriman, ed., *Oceanography: The Past* (New York: Springer-Verlag, 1980), 522–539.

10. William Stimpson, "Synopsis of the Marine Invertebrates of Grand Manan," *Smithsonian Contributions to Knowledge* 6 (1854): 6; Vasile, "Fine View from the Shore."

11. For an example of Baird's role in promoting and organizing scientific collecting, see Baird to Louis Agassiz, March 9, 1854, in Elmer Charles Herber, ed., *Correspondence between Spencer Fullerton Baird and Louis Agassiz: Two Pioneer American Naturalists* (Washington, D.C.: Smithsonian Institution, 1963). For Stimpson's teaching Baird about dredging, see Stimpson to Baird, Aug. 21, 1863, and Stimpson to Baird, Sept. 4, 1863, in SIA, SFB, RU 7002, Box 33.

12. Allen, *Naturalist in Britain,* 83–107; Jack Morrell and Arnold Thackray, *Gentlemen of Science: Early Years of the British Association for the Advancement of Science* (New York: Oxford University Press, 1981).

13. Philip Henry Gosse, *The Ocean* (London: John van Voorst, 1845); Gosse, *A Naturalist's Rambles on the Devonshire Coast* (London: John van Voorst, 1853). Later Gosse wrote *Tenby: A Sea-Side Holiday* (London: John van Voorst, 1859), and *The Wonders of the Great Deep; or the Physical, Animal, Geological, and Vegetable Curiosities of the Ocean, with an Account of the Submarine Explorations beneath the Sea, Diving, Ocean Telegraphing, Etc.* (Philadelphia: Quaker City Publishing House, 1874). Charles Kingsley, *Glaucus; or, The Wonders of the Shore* (Cambridge: Macmillan, 1855); Elizabeth Cary Agassiz and Alexander Agassiz, *Seaside Studies in Natural-History* (Boston: Ticknor & Fields, 1865). On popularization, see Bernard Lightman, "'The Voices of Nature': Popularizing Victorian Science," in Lightman, ed., *Victorian Science in Context* (Chicago: University of Chicago Press, 1997), 187–211.

14. Margaret Gatty, *Parables from Nature* (London, 1855); Suzanne Le-May Sheffield, *Revealing New Worlds: Three Victorian Women Naturalists* (London: Routledge, 2001), 13–74.

15. See Rehbock, "Victorian Aquarium," 522–539. Gosse also wrote *The Aquarium: An Unveiling of the Wonders of the Deep Sea* (London: John van Voorst, 1854).

16. Allen, *Naturalist in Britain,* 73–93; Peter Bailey, "Leisure, Culture, and the Historian: Reviewing the First Generation of Leisure Historiography in Britain," *Leisure Studies* 8 (1989): 107–127; Peter Bailey, *Leisure and Class in Victorian England: Rational Recreation and the Contest for Control, 1830–1885* (London: Methuen, 1987); Larsen, "Not Since Noah," 142–143; Thomson, *Depths of the Sea,* 280; George B. Goode to William H. Dall, July 17, 1876, SIA, WHD, RU 7073, Box 11, folder 1; Edward Forbes, *The Natural History of the European*

Seas, ed. and continued by Robert Godwin-Austen (London: John van Voorst, 1859), 10.

17. George S. Brady to Joshua Alder, July 9, 1865, NHM General Library, MSS ALD, vol. 2, 246.

18. Forbes, *Natural History of the European Seas,* 102; Dr. George Johnston to William Thompson, Sept. 13, 1840, and Oct. 10, 1840, in James Hardy, ed., *Selections from the Correspondence of Dr. George Johnston* (Edinburgh: David Douglas, 1892), 150–151, 156–157.

19. Forbes to Alder, July 5, [no year], NHM General Library, MSS ALD, vol. 3, 471; Stimpson to Baird, July 12, 1861, and Stimpson and Theodore Gill to Baird, March 11, 1861, SIA, SFB, RU 7002, Box 33. This sort of camaraderie also characterized midcentury fieldwork other than dredging. See James A. Secord, "The Geological Survey of Great Britain as a Research School, 1839–1855," *History of Science* 24 (1986): 223–275, and Hugh R. Slotten, *Patronage, Practice, and the Culture of American Science: Alexander Dallas Bache and the Coast Survey* (New York: Cambridge University Press, 1994).

20. Alain Corbin, *The Lure of the Sea: The Discovery of the Seaside in the Western World, 1750–1840,* trans. Jocelyn Phelps (Cambridge: Polity Press, 1994), 57–96. At midcentury, respectable working-class men, and even middle-class young men, embarked on whaling voyages for health reasons. Margaret S. Creighton, *Rites and Passages: The Experience of American Whaling, 1830–1870* (Cambridge: Cambridge University Press, 1995); Robert Foulke, "The Literature of Voyaging," in Patricia Ann Carlson, ed., *Literature and Lore of the Sea* (Amsterdam: Rodolphi, 1986), 1–73.

21. The "sea breeze" quote is from George B. Goode to William Haley Dall, Salem, MA, July 29, 1877, SIA, WHD, RU 7073, Box 11, folder 1. Goode spent several winters collecting and dredging in the Bahamas for his health. Forbes to Andrew C. Ramsey, Aug. [no year], and Sept. 24, 1846, ICA, A. C. Ramsey Papers, Correspondence from Forbes, 8/389/13–14; Forbes to Joseph Hooker, Thursday [n.d.], KEW, Letters to Joseph D. Hooker, vol. 8, 157.

22. Edward F. Rivinus and E. M. Youssef, *Spencer F. Baird of the Smithsonian* (Washington, D.C.: Smithsonian Institution Press, 1992); Thomas H. Huxley, "Journal," ICA, THH, Notebook, vol. 125; Adrian Desmond, *Huxley: From Devil's Disciple to Evolution's High Priest* (Reading, MA: Addison-Wesley, 1997), 212–215.

23. Thomson, *Depths of the Sea,* 267.

24. Carpenter to Huxley, Sept. 26, 1855, ICA, THH, vol. 12, 82–88.

25. John G. Jeffreys to Joshua Alder, May 24, 1864, NHM General Library, MSS ALD, vol. 4, 733.

26. Rev. Leonard Jenyns, "Report on Zoology," *BAAS Report* (1834), p. 143. For an overview of the British Association dredging activities, see A. L. Rice and J. B. Wilson, "The British Association Dredging Committee: A Brief History," in Sears and Merriman, ed., *Oceanography,* 373–385. Rehbock, "Early Dredgers"; Mills, "Edward Forbes, John Gwyn Jeffreys." See also O. J. R. Howarth, *The British Association for the Advancement of Science, 1831–1921* (London: Burlington House, 1922).

27. Philip F. Rehbock, *The Philosophical Naturalists: Themes in Early Nineteenth-*

Century British Biology (Madison: University of Wisconsin Press, 1983); Mary P. Winsor, *Starfish, Jellyfish, and the Order of Life: Issues in Nineteenth-Century Science* (New Haven: Yale University Press, 1976), 4–5, 175–178; Allen, *Naturalist in Britain;* Pamela M. Henson, "Evolution and Taxonomy: J. H. Comstock's Research School in Evolutionary Entomology at Cornell Unviersity, 1874–1930" (Ph.D. diss., University of Maryland, College Park, 1990), 24–91; Larsen, "Not Since Noah"; Paul Farber, "The Transformation of Natural History in the Nineteenth Century," *Journal of the History of Biology* 15 (1982): 145–152; Paul L. Farber, *Finding Order in Nature: The Naturalists' Tradition from Linneaus to E. O. Wilson* (Baltimore: Johns Hopkins University Press, 2000).

28. Eric Mills has characterized Jeffreys as more preoccupied by new species than by zoogeographic problems, in "Edward Forbes, John Gwyn Jeffreys, and British Dredging," 530. The same pattern of rooting theoretical issues firmly in the routine work of natural history holds true for Alfred Merle Norman; see Mills, "One 'Different Kind of Gentleman': Alfred Merle Norman (1831–1918), Invertebrate Zoologist," *Linnean Society of London, Zoological Journal* 68 (1980): 69–98.

29. Martin J. S. Rudwick coined the term "correspondence networks"; see Rudwick, *The Great Devonian Controversy: The Shaping of Scientific Knowledge among Gentlemanly Specialists* (Chicago: University of Chicago Press, 1985); Larsen, "Not Since Noah," 288–355.

30. Peter S. Davis, "Marine Biologists," in A. G. Lunn, ed., *A History of Naturalists in N.E. England* (Newcastle upon Tyne: University of Newcastle upon Tyne, 1983), 52–53. Alder's collection, and Norman's as well, were donated to the British Museum and are now at the Natural History Museum in London. The quote was by E. Ray Lankester, quoted in Mills, "One 'Different Kind of Gentleman,'" 69–98. Davis, "Marine Biologists," 55–58; Sidney F. Harmer, "Canon Alfred Merle Norman, FRS," *Nature* 102 (1918): 188–189; Anon, "Alfred Merle Norman, 1831–1918" *Proceedings of the Malacological Society of London* 13 (1919): 116–117; J. Cosmo Melvill, "Obituary Notice: The Rev. Canon Alfred Merle Norman, DCL, FRS," *Journal of Conchology* 16 (1919): 40–41.

31. G. R. Agassiz, ed., *Letters and Recollections of Alexander Agassiz with a Sketch of His Life and Work* (Boston: Houghton Mifflin, 1913), 168.

32. Rehbock, "Early Dredgers", 311–323. For an excellent study of stratigraphy and fossils, see Martin J. S. Rudwick, *The Meaning of Fossils: Episodes in the History of Palaeontology,* 2nd ed. (Chicago: University of Chicago Press, 1985).

33. Rehbock, "Early Dredgers," 318.

34. See the series of letters (1840–1851) between Forbes and William Thompson at The Manx Museum and National Trust, Isle of Man, Archives MS 2148A. "Memoir of the Late William Thompson, Esq.," in William Thompson, *The Natural History of Ireland,* vol. 4 (London: Reeve, Benham and Reeve, 1849–1856), x–xxx. *BAAS Report* (1849), p. 72. At the 1856 meeting, Ball also exhibited a dredge, "which he had found of the greatest use in making dredging excursions." "Transactions of the Sections, Zoology," *BAAS Report* (1856), p. 91. A. L. Rice and J. B. Wilson, "The British Association Dredging Committee: A Brief History," in Sears and Merriman, ed., *Oceanography,* 376.

35. Rice and Wilson, "British Association Dredging Committee," 373–385.

36. Joshua Alder and Albany Hancock, *On the British Nudibranchiate Mollusca* (London: Ray Society, 1845–55); Davis, "Marine Biologists," 45–63; Allen, *Naturalist in Britain,* 145–146.

37. Davis, "Marine Biologists," 49–56, 58.

38. First names or initials of dredgers are given if known.

39. Edward Waller to Alfred Merle Norman, June 2 and 22, 1864, NHM General Library, MSS ALD, 729–730. For other examples of hiring collectors, see Joshua Alder to Norman, Jan. 29, 1861, and June 10, 1863, NHM General Library, MSS ALD, vol. 1, 30–31, 80A.

40. Sidney Ross, "Scientist: The Story of a Word," *Annals of Science* 18 (1962): 65–85; Robert K. Merton, "De-Gendering Man of Science: The Genesis and Epicene Character of the Word Scientist" in *Sociological Visions,* ed. Kai Erikson (Lanham, MD: Rowman & Littlefield, 1997).

41. Forbes, *Natural History of the European Seas,* 1–2; the quote is from p. 4. John Gwyn Jeffreys, "Fourth Report on Dredging among the Shetland Isles," *BAAS Report* (1868), p. 431; Janet Browne, *The Secular Ark: Studies in the History of Biogeography* (New Haven: Yale University Press, 1983); Edward Forbes to John E. Gray, Oct. 13, 1839, APS, John Edward Gray Papers, B G784; Susan Faye Cannon characterizes Charles Darwin's work, and natural history aimed at answering biogeographical questions, as being influenced by Alexander von Humboldt's definition of science as the accurate, measured study of widespread but interconnected real phenomena, including the astronomy, physics, and biology of the earth, all viewed from a geographical perspective. Although naturalists did not adhere to the concern for accurate, quantitative measurement, the search for an underlying, uniform law was typically Humboldtian. This tradition was based on fieldwork, and thus the competent scientific traveler ought to cope as effectively with "the carelessness of clumsy donkeys" as with scientific measurement. See Cannon, *Science in Culture: The Early Victorian Period* (New York: Science History Publications, 1978), 86–92. Forbes used a system for data collecting to study biogeography, that is, making a record of all organisms collected in each haul on preprinted dredging papers, a practice he introduced to the BAAS dredging committees; see Mills, "Edward Forbes, John Gwyn Jeffreys," 512. Larsen, "Not Since Noah," 209.

42. Debra Lindsay, *Science in the Subarctic: Trappers, Traders, and the Smithsonian Institution* (Washington, D.C.: Smithsonian Institution Press, 1993), 27–40; Rivinus and Youssef, *Spencer F. Baird of the Smithsonian,* 120–121, 141–151; Dean C. Allard, *Spencer Fullerton Baird and the U.S. Fish Commission* (New York: Arno Press, 1978).

43. Rice and Wilson, "British Association Dredging Committee," 378.

44. Alder to Norman, June 8, 1859, Sept. 11, 1860, and Jan. 26, 1862, NHM General Library, MSS ALD, vol. 1, 14–15, 25, 60–61.

45. Thomson, *Depths of the Sea,* 241.

46. Histories of yachting were written around the turn of the century, focusing on particular yacht clubs, vessels, or races. The recent interest in sports history has produced some books on sports in general that briefly mention yachting. In these, the authors concentrate on racing and the high society surrounding yachting, rather than on ocean cruising. See J. D. Jerrold Kelley, *American Yachts:*

Their Clubs and Races (New York: C. Scribner's Sons, 1884), 2–5; Douglas Phillips-Birt, *Fore and Aft Sailing Craft and the Development of the Modern Yacht* (London: Seeley, Service and Co., 1962), 100; Peter Heaton, *Yachting: A History* (London: BT Batsford, 1955), 57–67, 124; Montague Guest and William B. Boulton, *The Royal Yacht Squadron: Memorials of Its Members, with an Inquiry into the History of Yachting* (London: John Murray, 1903), 1–3, 14–15, 28–31.

47. Guest and Boulton, *Royal Yacht Squadron*, 1–3, 28–31, 59, 169, 285–286; C. M. Gavin, *Royal Yachts* (London: Rich & Cowan, 1932), 91; Corbin, *The Lure of the Sea*, 253–274; Heaton, *Yachting*, 93–95, 103.

48. The quote is from Heaton, *Yachting*, 77; Guest and Boulton, *Royal Yacht Squadron*, 36–37, 75–77; Phillips-Birt, *Fore and Aft Sailing Craft*, 150. The Northern Yacht Club, founded in 1824, was organized in Belfast by gentlemen in northern Ireland and western Scotland. In 1827, the group that became the Western Yacht Club formed with members from England and Ireland and, in 1831, the Royal Irish Yacht Club was formed. In the 1830s, the Royal London, Royal Southern, and Royal Harwich Yacht Clubs organized yachting closer to the metropolis. Melvin Adelman, *A Sporting Time: New York City and the Rise of Modern Athletics, 1820–1870* (Urbana: University of Illinois Press, 1986), 201.

49. Guest and Boulton, *Royal Yacht Squadron*, 61, 231–233, 282; Gavin, *Royal Yachts*, 127–128; Winfield M. Thomson and Thomas W. Lawson, *The Lawson History of America's Cup: A Record of Fifty Years* (Boston: published privately by Thomas W. Lawson, 1902), 273–279.

50. The quote is from Guest and Boulton, *Royal Yacht Squadron*, 299; Adelman, *A Sporting Time*, 202; John Rickards Betts, *America's Sporting Heritage, 1850–1950* (Reading, MA: Addison-Wesley, 1974), 149–152; Bill Robinson, *Legendary Yachts* (New York: Macmillan, 1971), 5; Anon., "The Little Ship of the Atlantic," *The Illustrated London News,* Supplement, Sept. 15, 1866, p. 261.

51. Heaton, *Yachting*, 114–120, 159–163; Robinson, *Legendary Yachts*, 13. For comparison purposes, shooting preserved game, a very elite pastime, cost £500–600 a year; hunting, £100+, rowing, £10+, mountaineering, £50+, golf, £20–30, and cycling, £10. John Lowerson, *Sport and the English Middle Classes, 1879–1914* (Manchester: Manchester University Press, 1993), 13, 29–31, 52.

52. Guest and Boulton, *Royal Yacht Squadron*, 193; Kelley, *American Yachts*, 6; Lowerson, *Sport and the English Middle Classes*, 194; Adelman, *A Sporting Time*, 201.

53. Anne Brassey, *Around the World in the Yacht 'Sunbeam'* (New York: Henry Holt & Co., 1878), 128–131.

54. Guest and Boulton, *Royal Yacht Squadron*, 90–91, 137; Dr. Phipps, "On the Sailing Powers of Two Yachts Built on the Wave Principle," *BAAS Report* (1846), p. 112.

55. Kelley, *American Yachts*, 6; Gavin, *Royal Yachts*, 144.

56. Rehbock, "Early Dredgers," 314–318 *BAAS Report* (1839), p. xxvi; *Dictionary of National Biography* vol. 18, 467.

57. George C. Hyndman, "Note of Species Obtained by Deep Dredging Near Sana Island, off the Mull of Cantire," *BAAS Report* (1842), pp. 70–71; Hyndman, "Report of the Proceedings of the Belfast Dredging Committee," *BAAS Report* (1857), pp. 220–229.

58. The quote is from Forbes, *Natural History of the European Seas,* 106–107; see also 125–130; Richard MacAndrew, "Robert McAndrew, 1802–1873" (privately published, 2002); Anon., "Robert MacAndrew," *Proceedings of the Linnean Society of London* 18 (1873–74): xxxiii–xxv. MacAndrew's last name is sometimes spelled "McAndrew." William Herdman, *Founders of Oceanography and Their Work: An Introduction to the Science of the Sea* (London: Edward Arnold, 1923), 19; Rehbock, "Early Dredgers," 340–346.

59. The quotes are from Edward Forbes, "Report on the Investigation of British Marine Zoology by Means of the Dredge," *BAAS Report* (1850), pp. 192–263. *BAAS Report* (1845), p. xix; Robert MacAndrew and Edward Forbes, "Report of the Dredging Committee for 1844," *BAAS Report* (1844), pp. 63–64; Forbes, "On Some Animals New to the British Seas, Discovered by Mr. MacAndrew," *BAAS Report* (1844), p. 64; Rice and Wilson, "British Association Dredging Committee," 378.

60. Robert MacAndrew, "Notes on the Distribution and Range in Depth of Mollusca and Other Marine Animals Observed on the Coasts of Spain, Portugal, Barbary, Malta, and Southern Italy in 1849," *BAAS Report* (1850), pp. 264–304; MacAndrew, "Robert McAndrew," 15–19; Forbes, *Natural History of the European Seas,* 129–130. MacAndrew left his collection to the University of Cambridge.

61. The quote is from Forbes, *Natural History of the European Seas,* 106–107; *BAAS Reports* (1858), p. xi, (1859), p. xlix, (1860), p. xl, (1861), p. xl; Rice and Wilson, "British Association Dredging Committee," 378–379.

62. For biographical information on Jeffreys, see David Heppell, *Dictionary of Scientific Biography* vol. 7, 91–92; William B. Carpenter, "Obituary Notices of Fellows Deceased," *Proceedings of the Royal Society of London* 38 (1885): xiv–xviii; and Anon., "John Gwyn Jeffreys," *Nature* 31 (1885): 317–318. Mills, "Edward Forbes, John Gwyn Jeffreys," 509–511.

63. *BAAS Report* (1858), p.xl. Rehbock states that they were friends in "Early Dredgers," 346, but without citing a source. Mills, "Edward Forbes, John Gwyn Jeffreys," 527–532; MacAndrew, "Robert McAndrew," 36–39.

64. Mills, "Edward Forbes, John Gwyn Jeffreys," 530–531.

65. John G. Jeffreys, "Report of the Committee Appointed for Exploring the Coasts of Shetland by Means of the Dredge," *BAAS Report* (1863), pp. 70–74.

66. Jeffreys, "Report of the Committee Appointed for Exploring the Coasts," *BAAS Report* (1863), p. 73; Alfred Merle Norman, "Shetland Final Dredging Report, Part II. On the Crustacea, Tunicata, Polyzoa, Echinodermata, Actinozoa, Hydrozoa, and Porifera," *BAAS Report* (1868), p. 248.

67. Jeffreys to Norman, May 26, 1864 (quote is from here), and June 6, 1864, NHM General Library, MSS ALD, vol. 4, 735–737, 739–741; John G. Jeffreys, "Further Report on Shetland Dredgings," *BAAS Report* (1864), p. 327; Jeffreys, "Last Report on Dredging among the Shetland Isles," *BAAS Report* (1868), p. 251.

68. Jeffreys to Alder, Aug. 3, 1844, NHM General Library, MSS ALD, vol. 4, 697; Jeffreys to Alder, June 30, 1861, and May 24, 1864, and Jeffreys to Norman, May 26, 1864, NHM General Library, MSS ALD, vol. 4, 728, 733, 735–737.

69. Jeffreys to Alder, July 28, 1861, NHM General Library, MSS ALD, vol. 4, 726–727.

70. Jeffreys to Norman, July 2, 1861, and June 6, 1864 (quote is from here), NHM General Library, MSS ALD, vol. 4, 729–730, 734–741.

71. Jeffreys to Alder, June 30, 1861, and Jeffreys to Norman, May 26, 1864, and June 6, 1864, NHM General Library, MSS ALD, vol. 4, 728, 735–737, 739–741; Thomson, *Depths of the Sea,* 85. Again in 1871, Laughrin sailed on HMS *Shearwater* to the Mediterranean on a scientific cruise led by William B. Carpenter. Carpenter to George G. Stokes, June 28, 1871, RSA, MC 9.227; Carpenter, "Report on Scientific Researches Carried on during the Months of August, September, and October, 1871, in HM Surveying-Ship 'Shearwater,'" *Proceedings of the Royal Society of London* 20, no. 138 (June 1872): 535–644.

72. Heaton, *Yachting,* 95; Gavin, *Royal Yachts,* 102, 198–199; Brassey, *Around the World in the Yacht "Sunbeam.";* Lowerson, *Sport and the English Middle Classes,* 211. A long tradition existed of women accompanying their captain-husbands to sea in whaling and merchant vessels, particularly on long voyages. See Lisa Norling, *Captain Ahab Had a Wife: New England Women and the Whalefishery, 1720–1870* (Chapel Hill: University of North Carolina Press, 2000).

73. "The British Association Meeting at Edinburgh," *Nature* 4 (1871): 313–315; W. R. Hughes, "The Recent Marine Excursion Made by the Society to Teignmouth," *Nature* 9(1873–74): 253–254.

74. The "diligent" quote is from Forbes, *Natural History of the European Seas,* 78–79. Marcia Myers Bonta, *Women in the Field: America's Pioneering Women Naturalists* (College Station: Texas A & M University Press, 1991); Susan Drain, "Marine Botany in the Nineteenth Century: Margaret Gatty, the Lady Amateurs, and the Professionals," in *Victorian Studies Association Newsletter* 53 (Spring 1994): 7; Sheffield, *Revealing New Worlds,* 31. The Johnston quote is from Davis, "Marine Biologists," 47–48. Elizabeth Platts, "In Celebration of the Ray Society, Established 1844, and Its Founder George Johnston (1797–1855)" (London: Ray Society Publication, 1995).

75. John G. Jeffreys, "Presidential Address," *Transactions of the Hertforshire Natural History Society,* 1, pt. 1 (Sept. 1880): 1–4; John Gwyn Jeffreys, *British Conchology,* vol. 1 (London: John van Voorst, 1852), viii; Mills, "Edward Forbes, John Gwyn Jeffreys," 511.

76. Sally G. Kohlstedt, "In From the Periphery: American Women in Science, 1830–1880," *Signs* 4 (1978): 81–96; Larsen, "Not Since Noah," 58–128. As Larsen points out, not all arguments against excessive destruction of animals and plants were aimed at feminine sensibilities. Jeffreys advised his "brother naturalists to be moderate in their captures of animals and plants" in his presidential address to the Hertforshire Natural History Society, cited in the previous note. See also Anne B. Shteir, "Linnaeus' Daughters: Women and British Botany," in Barbara J. Harris and JoAnne K. McNamara, ed., *Women and the Structure of Society* (Durham: Duke University Press, 1984). Rehbock, "Victorian Aquarium," 351–355.

77. Davis, "Marine Biologists," 94–96.

78. Earlier, Louis Agassiz and his family had run a school for girls that included instruction in natural history. Edward Lurie, *Louis Agassiz: A Life in Science* (Baltimore: Johns Hopkins University Press, 1988), 200–202; Margaret W. Rossiter,

Women Scientists in America: Before Affirmative Action, 1940–1972 (Baltimore: Johns Hopkins University Press, 1995), 86–89.

79. For women, barriers to oceanography remained high until quite recently. One Scottish zoologist, Sheina Marshall, managed to participate in the Great Barrier Reef Expedition of 1928–29 and other inshore fieldwork only because the vessel returned to port at night. A small number of women contributed to oceanography by analyzing data on shore, especially during World War II. Only one woman earned a doctorate in oceanography between 1948 and 1961. Even women scientists and graduate students working at the Scripps Oceanographic Institution and the Woods Hole Oceanographic Institution were categorically barred from going to sea until the 1970s. So strict was the prohibition that a student who stowed away on a research vessel in the mid-1950s, in the hopes of being allowed to remain aboard for the cruise, was immediately beached and reprimanded. Ben McKelway, "Women in Oceanography," *Oceanus* 25, no.4 (1982–83): 77–79; F. S. Russell, "Sheina Macalister Marshall, 20 April 1896–7 April 1977," *Biographical Memoirs of Fellows of the Royal Society*, 24 (November 1978): 369–389; Naomi Oreskes, "Looking For a Few Good Women: The Bathythermograph and Military Patronage of Feminized Scientific Labor," paper delivered at the History of Science Society Annual Meeting, November 1995; Oreskes, "Objectivity or Heroism: On the Invisibility of Women in Science," in Henrika Kuklick and Robert E. Kohler, ed., *Science in the Field, Osiris* 11 (1996): 87–116; Kathleen Broome Williams, *Improbable Warriors: Women Scientists and the U.S. Navy in World War II* (Annapolis: Naval Institute Press, 2001); Kathleen Crane, *Sea Legs: Tales of a Woman Oceanographer* (Cambridge, MA: Westview Press, 2003); Rossiter, *Women Scientists in America*, 58–59, 80, 86–89.

5. Dredging the Moon

1. John Gwyn Jeffreys and William B. Carpenter, *The Valorous Expedition: Papers by Dr. Gwyn Jeffreys, FRS, and Dr. Carpenter, CB, FRS* (London: Taylor & Francis, 1876), HO, Hydrographic Publication 60.

2. George Barlee to Joshua Alder, Dec. 2, 1856, NHM General Library, MSS ALD, vol. 1, 172.

3. Indeed, Forbes's contemporaries insisted that, had he not died at the tragically young age of thirty-eight, he would have been the first to repudiate his own theory. Forbes's champions attributed his mistaken belief to imperfect collecting techniques. They also pointed out that life is indeed rare below 300 fathoms in the Mediterranean, where Forbes conducted his deepest dredgings. Charles Wyville Thomson, *The Depths of the Sea: An Account of the General Results of the Dredging Cruises of HMSS "Porcupine" and "Lightning" during the Summers of 1868, 1869, and 1870* (New Yorkn: Macmillan, 1873), 18; George C. Wallich, Notebook, "On the History of Deep-Sea Exploration (#2)," Wellcome Institute Library, London, MS 4,963.

4. See, for example, Alison Winter, "'Compasses All Awry': The Iron Ship and the Ambiguities of Cultural Authority in Victorian Britain," *Victorian Studies* 37 (Autumn 1994): 69–98.

5. A. L. Rice, Harold L. Burstyn, and A. G. E. Jones, "G. C. Wallich, MD: Megalo-

maniac or Mis-used Oceanographic Genius?" *Journal of the Society for the Bibliography of Natural History* 7, no. 4 (1976): 437–439.

6. Margaret Deacon, *Scientists and the Sea, 1650–1900: A Study of Marine Science,* 2nd ed. (Brookfield, VT: Ashgate Publishing Co., 1997), 228–229, 281–283, 309.

7. James Clark Ross, *A Voyage of Discovery and Research in the Southern and Antarctic Regions during the Years 1839–1843* (London: John Murray, 1847).

8. Edward Forbes, *The Natural History of the European Seas,* ed. and continued by Robert Godwin-Austen (London: John van Voorst, 1859).

9. Anne L. Larsen, "Not Since Noah: The English Scientific Zoologists and the Craft of Collecting, 1800–1840" (Ph.D. diss., Princeton University, 1993), 226–287; Martin Rudwick, *The Great Devonian Controversy: The Shaping of Scientific Knowledge among Gentlemanly Specialists* (Chicago: University of Chicago Press, 1985).

10. Chandros Michael Brown, "A Natural History of the Gloucester Sea Serpent: Knowledge, Power, and Culture of Science in Antebellum America," *American Quarterly* 42, no. 3 (1990): 402–436.

11. William Stimpson, "Journal of a Cruise in the US Ship Vincennes to the North Pacific Expedition, Behring Straits, Etc.," SIA, NPEE, RU 7253, Box 1; Sherrie Lyons, "Sea Monsters: Myth or Genuine Relic of the Past," in Keith R. Benson and Philip F. Rehbock, ed., *Oceanographic History: The Pacific and Beyond* (Seattle: University of Washington Press, 2002); Richard Ellis, *Monsters of the Sea: The History, Natural History, and Mythology of the Ocean's Most Fantastic Creatures* (New York: Alfred A. Knopf, 1995).

12. Spencer F. Baird to George Brown Goode, Newport, RI, Sept. 24, 1880 (the "puzzle" quote is from here), and Sept. 29, 1880, SIA, SFB, RU 7002, Box 21. The "basking" quote is from Spencer F. Baird, "Assistance to Commission of Fish and Fisheries," Circular to S. I. Kimball, General Superintendent, Life-Saving Service, Washington, D.C., Treasury Department, Feb. 2, 1883, SIA, Rhees Papers, RU 7081, Box 15. Louis Agassiz to Baird, Nov. 23, 1868, and Baird to Agassiz, Nov. 25, 1868, in Elmer Charles Herber, ed., *Correspondence between Spencer Fullerton Baird and Louis Agassiz: Two Pioneer American Naturalists* (Washington, D.C.: Smithsonian Institution, 1963).

13. After Ross's death, the collections were rediscovered, but in a state totally unfit for study. Charles Wyville Thomson and John Murray, ed., *Report of the Scientific Results of the Exploring Voyage of the HMS "Challenger," 1873–76: Narrative of the Cruise,* vol. 1 (London: HMSO, 1885), xliii.

14. George C. Wallich, "Daily Diary of Voyage of 'Bulldog,,'" July 4, 12 (quote is from here), 20, Aug. 10, 14, 27, 1860, NHM General Library, MSS WAL; hereafter cited as "Wallich Journal."

15. Leopold McClintock, "Atlantic Soundings by HMS Bulldog, 1860," 15, HO, C. Original Document Series, 390.

16. The quotes are from "Wallich Journal," Oct. 15, 16, 1860; George Wallich, *The North Atlantic Sea-Bed: Comprising a Diary of the Voyage on Board HMS "Bulldog" in 1860* (London: John van Voorst, 1862), 68–69.

17. Wallich, *North Atlantic Sea-Bed,* 68–69; Thomson, *Depths of the Sea,* 25; Rice, Burstyn, and Jones, "G. C. Wallich," 433–434.

18. The letter is quoted in "Wallich Journal," Sept. 9, 1860; Thomson, *Depths of the Sea,* 27–28.

19. Thomson, *Depths of the Sea,* 29, 272 (the quote is from here).

20. Eric L. Mills notes that Edward Forbes's book was more popular than his earlier work on the historical biogeography of the British Isles, but less scientifically convincing; see Mills, "Edward Forbes, John Gwyn Jeffreys, and British Dredging before the *Challenger* Expedition," *Journal of the Society for the Bibliography of Natural History* 8, no. 4 (1978): 518–519.

21. P. P. Carpenter to William H. Dall, June 5, 1870, SIA, WHD, RU 7073, Box 9, f. 1. Philip P. Carpenter was the brother of William B. Carpenter. Many zoologists consulted private collections like Jeffreys's in lieu of museum collections, which were usually not as extensive or as well cared for as private collections until later in the century. Edward Forbes frequently visited Jeffreys's collection. Forbes to Andrew S. Ramsey, Sept. 14, 1845, Wexford, ICA, ACR, Correspondence from Forbes, 8/389/2; Forbes to Alder, Edinburgh [1836?], NHM General Library, MSS ALD, vol. 3, 470. For an assessment of the conditions of museum collections at the time, see David Allen, *The Naturalist in Britain: A Social History* (Princeton: Princeton University Press, 1994), 75. Mills, "Edward Forbes and John Gwyn Jeffreys," 509–511.

22. Jeffreys to Wallich, March 11, 1866, NHM General Library, MSS WAL. Wallich had earlier sent Jeffreys some shells to identify from the dredgings from HMS *Bulldog* in 1860. Wallich, "Report on the Soundings Taken on Board HMS 'Bulldog,' between July 1 and November 8, 1860," 140, NHM, Mineralogy Library, John Murray Library; Jeffreys to Charles Lyell, Oct. 20, 1868, EUL, Gen. 112, 2956; Jeffreys to O. A. L. Mörch, Dec. 27, 1862, July 30, 1865, and Sept. 25, 1865, July 10, 1871, NHM Zoology Library 89, f J, John Gyn Jeffreys Collection, 2, 22–24, 87; George Goode to William Dall, Richmond, July 1, 1883, SIA, WHD, RU 7073, Box 11, f. 1; Dall to Jeffreys, Oct. 6, 1872, RSA, MC 9.414; John G. Jeffreys, *British Conchology,* vol 2 (London: John van Voorst, 1863), iii; William A. Deiss and Raymond B. Manning, "The Fate of the Invertebrate Collections of the North Pacific Exploring Expedition, 1853–1856," in Alwyne Wheeler and James H. Price, ed., *History in the Service of Systematics* (London: Society for the Bibliography of Natural History, 1981), 79–85.

23. William Chimmo to Hydrographer George Richards, HMS *Nassau,* April 27 [1870], HO, Surveyors Letters, or "S" Papers, 1870; William Benjamin Carpenter, *Introduction to the Study of Foraminifera* (London: Ray Society, 1862).

24. Carpenter to Thomas H. Huxley, Sept. 26, 1855, ICA, THH Papers, 12: 82–88.

25. Carpenter to Huxley, Sept. 26 and Oct. 2, 1855, ICA, THH Papers, vol. 12, 28–88, 89–90; Huxley to Dr. Dyster, Feb. 13, 1855, and April 9, 1855, in Leonard Huxley, *Life and Letters* (New York: D. Appleton and Co., 1900), vol. 1, 123–124, 125, 130; James Secord, "The Geological Survey of Great Britain as a Research School, 1839–1855," *History of Science* 24 (1986): 225–275.

26. Carpenter to Huxley, Sept. 26, 1855 (italics in the original). The Scottish dredging committee worked for three summers, reporting in 1856 specifically on their Lamlash collecting. Charles Popham Miles, "Dredging. Firth of Clyde. 1856," *BAAS Report* (1856), pp. 47–53.

27. Carpenter to Huxley, Sept 26, 1855.

28. Carpenter to Huxley, Sept. 26 and Oct. 2, 1855.

29. Carpenter to Huxley, Oct. 2, 1855; J. Estlin Carpenter, "Introductory Memoir," in William B. Carpenter, *Man and Nature: Essays Scientific and Philosophical*

(London: Kegan Paul, Trench 1888), 77; Barlee to Alder, Dec. 2, 1856; Joshua Alder to Alfred M. Norman, Jan. 1, 1857, NHM General Library, MSS ALD, vol. 1.

30. Carpenter, "Introductory Memoir," 88–91.

31. George C. Hyndman, "Report of the Belfast Dredging Committee for 1859," *BAAS Reports* (1859), pp. 116–119; *BAAS Reports* (1862), p. xli.

32. Charles Wyville Thomson to William B. Carpenter, May 30, 1868, RSA, MC 8.202; Alder to Norman, April, 15, May 8, Aug. 5, 1861, and March 2, 1863, NHM General Library, MSS ALD, vol. 1, 36, 38–39, 47–48, 73–74; Jeffreys to Alder, July 28, 1861, NHM General Library, MSS ALD, vol. 4, 726–727; Jeffreys to O. A. L. Mörch, April 24 and May 18, 1866, NHM Zoology Library 89, f J, John Gwyn Jeffreys Collection, 33, 37; Charles Darwin, *On the Origin of Species by Means of Natural Selection* (London: John Murray, 1859). Before Sars, Peder Christian Asbjørnsen, later a well-known folklorist, dredged the deep water of Hardangerfjord in 1853. His discovery of an unusual starfish, *Brisinga*, prompted more marine exploration in Norway. Liv Bliksrud, Geir Hestmark, and Tarald Rasmussen, *Vitenskapens utfordringer, 1850–1920* [The Challenges of Science, 1850–1920], Norsk idéhistorie [Norwegian History of Ideas], vol. 4 (Oslo: Aschoug, 2002), 23–69.

33. Thomson to Carpenter, May 30, 1868.

34. Robert Dawson to Aldred M. Norman, Nov. 22, 1864, NHM General Library, MSS ALD, vol. 2.

35. J. E. Davis, "Notes on Deep-Sea Sounding," (London: Hydrographic Office, Admiralty, 1867), HO, Hydrographic Department Publication 22, 11; Captain R. Wauchope, RN, "Meteorological and Hydrological Notes," *Memoirs of the Wernerian Natural History Society* 4 (1822): 161–172; William Chimmo, "Soundings and Temperatures of HMS 'Gannett,' North Atlantic," (1868), HO, Original Document Series, 401; Davis, "Soundings by HMS Cordelia, 1868," Sounding 10, p. 15.

36. William B. Carpenter to Professor George G. Stokes, June 15, 1871, RSA, MC 9.216; G. S. Ritchie, *The Admiralty Chart: British Naval Hydrography in the Nineteenth Century* (London: Hollis & Carter, 1967), 284–285, 313–327; Anita McConnell, *No Sea Too Deep: The History of Oceanographic Instruments* (Bristol: Adam Hilger, 1982). After retiring in 1874, Richards became managing director of the Telegraph Construction and Maintenance Company. J. B. Morrell argues that in the nineteenth century scientists sought to demonstrate their professionalism by attracting public recognition of themselves and their work; see Morrell, "The Patronage of Science in Nineteenth-Century Manchester," in G. L. E. Turner, ed., *The Patronage of Science in the Nineteenth Century* (Leyden: Noordhoff International Publishing, 1976), 53–94. Hugh R. Slotten notes the many tactics Bache used to gain public recognition and acceptance for Coast Survey work; see "The Dilemmas of Science in the United States: Alexander Dallas Bache and the U.S. Coast Survey," *Isis* 84, no. 1 (1993): 26–49.

37. William Chimmo to Hydrographer, Oct. 11, 1866, HO, Surveyors Letters, or "S" Papers, 1866/40.

38. Richard Hoskyn, "Atlantic Soundings by HMS Porcupine," HO, C. Original Document Series, 393; Davis, "Notes on Deep-Sea Sounding," 4–5.

39. Shortland, "The Hydra's Sounding Voyage, 1868," HO, Hydrographic Department Publication 27; McConnell, *No Sea Too Deep,* 58; Lieutenant-Commander E. K. Calver, "HMS Porcupine's Atlantic Journal, 1869," HO, C. Original Document Series, 394, 33.

40. Calver, "HMS Porcupine's Atlantic Journal, 1869," 35; Shortland, "The Hydra's Sounding Voyage, 1868," 11, 39–48.

41. Shortland, "The Hydra's Sounding Voyage, 1868." Reports and letters do not claim that Shortland was the first to think of applying the accumulator to deep-sea sounding, but they do not name another inventor.

42. The quote is from V. F. Johnson, "Remarks Explaining the Fittings Used and the Mode of Obtaining Deep Sea Soundings between the English and Newfoundland Banks, during the Months of April and May 1869," Screw Steamer "Greenwood," London, May 27, 1869, HO, Hydrographic Department Publication 29; Thomson, *Depths of the Sea,* 223–224.

43. Carpenter to Sabine, June 18, 1868, RSA, MC 8.222; Thomson, *Depths of the Sea,* 246.

44. Thomson to Carpenter, May 30, 1868.

45. Carpenter to Sabine, June 18, 1868.

46. The quotes are from Thomson, *Depths of the Sea,* 57. Carpenter to Norman, Aug. 10, 1868, Sept. 22, 1868, NHM General Library, MSS ALD, vol. 2, 325, 326. The *Lightning* was built in 1823 and entered in the Navy List in 1828 along with its sister ship, *Meteor,* as a steam-powered warship. They were not, however, the first powered warships. The *Comet* was built in 1822, but did not appear in the Navy List until 1831. Tony Rice, *British Oceanographic Vessels, 1800–1950* (London: The Ray Society, 1986), 97–100.

47. Carpenter to Norman, Sept. 22, 1868.

48. May, "Atlantic Soundings by HMS Lightning, Etc.," 6–8. The "special work" quote is from Thomson, *Depths of the Sea,* 246–247.

49. William B. Carpenter, John G. Jeffreys, and Charles Wyville Thomson, "Preliminary Report of the Scientific Exploration of the Deep Sea in HM Surveying Vessel 'Porcupine,' during the Summer of 1869" (London: Taylor & Francis, 1870), 397–398.

50. The quote is from Calver, "HMS Porcupine's Atlantic Journal, 1869," 1. W. G. Romaine to President of the Royal Society, Admiralty, March 19, 1869, RSA, MC 8.37; Carpenter, Jeffreys, and Thomson, "Preliminary Report," 401–403; Thomson, *Depths of the Sea,* 84.

51. Carpenter developed a theory that deep oceanic circulation was driven by density differences between the relatively warm water in the upper part of the water column (but below the surface layer) and the very cold deep water. At the time, neither Thomson nor Jeffreys accepted his ideas, nor did many of the scientists who participated in meteorological debates about oceanic circulation. Although Carpenter's interpretation was flawed, his basic idea about density-driven circulation has proved correct. Deacon, *Scientists and the Sea,* 320–328, 344, 348; Crosbie Smith and Norton Wise, *Energy and Empire: A Biography of Lord Kelvin* (Cambridge: Cambridge University Press, 1989).

52. The "no less" quote is from Calver, "HMS Porcupine's Atlantic Journal, 1869," 2. Jeffreys to Sabine, June 7, 1869, RSA, Sa.727; Thomson, *Depths of the Sea,* 90; Carpenter, Jeffreys, and Thomson, "Preliminary Report," 423.

53. Thomson, *Depths of the Sea,* 85–86; Calver to Richards, July 1, 1870, HO, Surveyors Letters, or "S" Papers, 1870.
54. Jeffreys to Sabine, June 7, 1869; Edward K. Calver to Hydrographer, Aug. 29, 1853, HO, Surveyors Letters, or "S" Papers, 1869, Calver; Peter Davis, "The Captain of the 'Porcupine': Edward Killiwick Calver, RN, FRS," *Porcupine Newsletter* 2, no. 6 (1982): 143–147; and Davis, "Collections of Material from the Dredging Expeditions of HMS Porcupine in NE England," *Porcupine Newsletter* 1 (1980): 159–161.
55. Jeffreys to Sabine, June 7, 1869; Calver, "HMS Porcupine's Atlantic Journal, 1869," 2 (the quote is from here); Thomson, *Depths of the Sea,* 83.
56. The quotes are from Calver to Richards, June 5, 1869, HO, Surveyors Letters, 46, Calver, 1858–1865; Calver, "HMS Porcupine's Atlantic Journal, 1869," 30.
57. Calver, "HMS Porcupine's Atlantic Journal, 1869," 40.
58. The "fair" and "idea" quotes are from Calver, "HMS Porcupine's Atlantic Journal, 1869," 39–41. Other quotes are from Thomson, *Depths of the Sea,* 105, 255–258. Tangles also kept sails from chafing against rigging.
59. Thomson, *Depths of the Sea,* 92–100.
60. Thomson to Norman, Aug. 7, 1869, reprinted in *BAAS Reports* (1869), p. 115.
61. The quote is from Carpenter, Jeffreys, and Thomson, "Preliminary Report," 449. William B. Carpenter, "On the Temperature and Animal Life of the Deep Sea," *Proceedings of the Royal Institution* 5 (1870): 1–21.
62. Vernon Lushington to W. Sharpey, Esq., Secretary of the Royal Society, Admiralty, May 10, 1870, RSA, MC 9.72; Thomson, *Depths of the Sea,* 178.
63. The quote is from the Royal Society, "Miscellaneous Committees, 1869–1884," notes of April 9, 1870 meeting, p. 5, RSA, CMB 2; Thomson, *Depths of the Sea,* 202–204; Carpenter to Sabine, Nov. 3, 1869, RSA, Sa.301; Deacon, *Scientists and the Sea,* 318.
64. Thomson, *Depths of the Sea,* 49.
65. The quotes are from Wolley to Secretary of the Royal Society, Admiralty, March 2, 1872, RSA, MC 9.334. Carpenter to Sabine, Nov. 3 and Nov. 6, 1869, RSA, Sa.301, Sa.302; Carpenter to Stokes, June 15, 1871.
66. Carpenter to Sabine, Nov. 3, 1869; Harold L. Burstyn, "Science and the Government in the Nineteenth Century: The Challenger Expedition and Its Report," *Bulletin de l' Institut Océanographique, Monaco,* special no. 2 (1968), vol. 2, 603–611.
67. William B. Carpenter, "Report on Scientific Researches Carried on during the Months of August, September, and October 1871, in HM Surveying-ship 'Shearwater,'" *Proceedings of the Royal Society* 20, no. 138 (1872): 535–644.
68. Carpenter to Sabine, Nov. 3, 1869. Royal Society, "Miscellaneous Committees, 1869–1884," notes of Nov. 10, 1871 meeting, p. 24, RSA, CMB 2; Maurice Yonge, "The Inception and Significance of the *Challenger* Expedition," *Proceedings of the Royal Society of Edinburgh* B 72, no. 1 (1971–72): 1–13.
69. Carpenter to Sabine, Nov. 3, 1869; Elisabeth Ross Micheli, "The British Empire Lays Claim to the Undersea World: The *Challenger* Expedition, 1872–1876" (M.Phil. thesis, Cambridge University, 1988), 17.
70. Carpenter to Sir John Herschel, May 6, 1868, RSA, Herschel Papers, Letters, 5.198; Jan. 24, 1869, RSA, HS 5.196. They also corresponded about temperature

and conditions of life at great depths. Matthew Fontaine Maury to Herschel, June 10, 1859, and Herschel to Maury, July 31, 1859, RSA, HS 12.308–309. Charles Lyell to Herschel, July 11, 1868, RSA, Herschel Papers, HS 11.445. John F. W. Herschel, ed., *A Manual of Scientific Inquiry; Prepared for the Use of Officers in Her Majesty's Navy; and Travelers in General* (London: John Murray, 1849).

71. Carpenter to Lyell, Dec. 12, 1872, APS, Darwin-Lyell Papers, B D25, L1; Carpenter to Lyell, Oct. 26, Nov. 13, 1871, and Feb. 25, 1872, EUL, Gen. 109, ff. 509, 513, 517; Jeffreys to Lyell, Oct. 20, 1868, EUL, Gen. 112, ff. 2956; Calver to Richards, Nov. 25, 1870, HO, Surveyors Letters, or"S" Papers, 161/1870; Carpenter to Sabine, Nov. 12, 1870, RSA, Sa.305.

72. Carpenter to Stokes, June 15, 1871; Royal Society, "Miscellaneous Committees, 1869–1874," p. 5, RSA, CMB 2; Deacon, *Scientists and the Sea*, 311.

73. Joseph Dayman, "Atlantic Soundings, Cyclops, 1857" (London, 1858), 63–68, HO, C. Original Document Series, 391; Adrian Desmond, *Huxley: From Devil's Disciple to Evoluation's High Priest* (Reading, MA: Addison-Wesley, 1994); Philip F. Rehbock, "Huxley, Haeckel, and the Oceanographers: The Case of *Bathybius haeckelii*," Isis 66 (1975): 504–533.

74. George Campbell, *Log-Letters from the "Challenger"* (London: Macmillan, 1877), 413; George Wallich to Joseph Hooker, June 1, 1860, KEW, English Letters, (1857–1900), vol. 104, 348; Great Britain, Submarine Telegraph Company, *Report of the Joint Committee Appointed by the Lords of the Committee of Privy Council for Trade and the Atlantic Telegraph Company to Inquire into the Construction of Submarine Telegraph; Together with the Minutes of Evidence and Appendix* (London: George Eyre and William Spottiswoode for HMSO, 1861), Jan. 13, 1860, p. 182.

75. Carpenter to Hooker, Aug. 6 and Aug. 11, 1847, and Nov. 13, 1869 (the quote is from here), KEW, Letters to J. D. Hooker, vol. 4, 13–14, 21; Wallich to Hooker, May 6, 1875, and Sept. 29, 1879, KEW, Letters to J. D. Hooker, vol. 21, 41–43, 46; Thomson to Hooker, Jan. 31 [1873], KEW, Voyage of HMS Challenger, Letters, Etc., p. 152 (this volume contains about half a dozen letters from Thomson discussing Moseley's collecting); Carpenter to Mr. President, Dr. Hooker, Jan. 21, 1875, and Jeffreys to President of the Royal Society [Hooker], Oct. 8, 1875, RSA, MC 10.200, 287.

76. William Thomson, "Address on the Forces Concerned in the Laying and Lifting of Deep-Sea Cables," *Proceedings of the Royal Society of Edinburgh* 5, no. 69 (1865–66): 595–605; Smith and Wise, *Energy and Empire*, 649–684, 723–753; McConnell, *No Sea Too Deep*, 60, 68–72. The navigational sounder sold for £20 while the deep-sea machine cost £127 10s.; "Price List. Sir William Thomson's Improved Mariner's Compass, (patent). Sir William Thomson's Improved Sounding Machine, (patent)," with stamp of F. M. Moore, Chronometer, Watch Maker, Optician, &c, Dublin. P. Morrison Swan to William Bottomley, Jan. 15, 1878, CUL, Add 7342 (Kelvin Papers), CS 439 and 108.

77. Before the expedition sailed, Rudolph von Willemöes-Suhm replaced William Stirling, who had been originally chosen by the Society. Von Willemöes-Suhm died during the expedition, while *Challenger* was in the Pacific.

78. For overviews of the *Challenger* expedition, see the official narrative, Thomson

and Murray, ed., *Report of the Scientific Results of the Challenger;* Deacon, *Scientists and the Sea,* 333–365; Yonge, "The Inception and Significance of the *Challenger.*"

79. State funding of science during the nineteenth century was proffered in response to threats to national security or supremacy. Morrell, "The Patronage of Science in Nineteenth-Century Manchester," 55–56.

80. The "remote past" quote is from William B. Carpenter to President of the Royal Society [1877], RSA, MC 11.143. John G. Jeffreys, "Report of the Committee Appointed for Exploring the Coasts of Shetland," 73; Deacon, *Scientists and the Sea,* 334; Alex S. Pang, "The Social Event of the Season: Solar Eclipse Expeditions and Victorian Culture," *Isis* 84, no. 2 (1993): 252–277.

81. John Gwyn Jeffreys, "The Deep-Sea Dredging Expedition in HMS 'Porcupine,'" *Nature* 1(1869–70): 166–168.

82. Anonymous, "Dredging the Gulf Steam," *Nature* 4 (1871): 87; Hugh R. Slotten, *Patronage, Practice, and the Culture of American Science: Alexander Dallas Bache and the Coast Survey* (New York: Cambridge University Press, 1994), 134–136; Edward Lurie, *Louis Agassiz: A Life in Science* (Baltimore: Johns Hopkins University Press, 1998), 131, 178, 371–377.

83. William Carpenter, "Notes," *Nature* 4 (1871): 107; David Forbes, "The Depths of the Sea," *Nature* 1 (1869–70): 100–101; Carpenter to Sabine, Nov. 3, 1869. For a discussion of the British perception of state support of science and the realities of state patronage in Germany, see Turner's "Introduction," and Brock, "The Spectrum of Science Patronage," in Turner, ed., *Patronage of Science,* 1–8, 173–206.

84. I agree with Margaret Deacon that the characterization of the *Challenger* expedition as the origin of oceanography greatly overemphasizes its significance; see *Scientists and the Sea,* 336, 368–370. McConnell, *No Sea Too Deep,* 71, 108, 113.

85. Henry H. Moseley, *Notes by a Naturalist: An Account of Observations Made during the Voyage of HMS "Challenger" round the World in the Years 1872–1876* (London: John Murray, 1892), 498–499; Charles Wyville Thomson, *Voyage of the Challenger: The Atlantic,* 2 vols. (London: Macmillan, 1877), vol. 1, 36, 57; John Murray, "HMS 'Challenger' Diary of Sir John Murray," Jan. 13, 1874, Jan. 19, 20, May 12, Sept. 4, 16, Oct. 9, 16, 1875, NHM, Mineralogy Library, John Murray Library, Section 1, vol. 1.

86. The lines of poetry are from Chief Engineer James H. Fergeson, "A letter from Jack Staylight to his old Shipmate 'Bill,'" copied into Pelham Aldrich [*Challenger* Journal], Jan. 16, 1873, RGS, LBR. MSS. AR 114A; John James Wild, "HMS 'Challenger': Diary of Mr. J. J. Wild," vol. 1, May 27, 1873, NHM, Mineralogy Library, John Murray Library, Acc. no. 179712, Section 1, no. 6–9 (Wild was the artist); Navigating Lieutenant Tizard et al. [Challenger Expedition Remark Book], vol. 1, Jan. 16, 1873, HO, C. Original Document Series, 276; Thomson, *Voyage,* 55.

87. The first quote is from George E. Belknap, "Deep-Sea Soundings in the North Pacific Ocean Obtained in the U.S. Steamer Tuscarora" (Washington, D.C.: Government Printing Office, 1874), 3–4. The second and third quotes are from Belknap to William Healey Dall, Aug. 5, 1874, SIA, WHD, RU 7073, Box 8, f. 5. McConnell, *No Sea Too Deep,* 61.

88. Dall to Spencer Baird, Oct. 30, 1872, SIA, WHD, RU 7073, Box 4. The "com-

promise" quote is from Dall to Julius Hilgard, Nov. 20, 1873, SIA, WHD, RU 7073, Box 4. William H. Dall, "Biographical Memoranda: Willian Cranch Healey Dall" (1926), SIA, WHD, RU 7073, Box 1, 82–83; George Belknap to Dall, Nov. 13, 1874, SIA, WHD, RU 7073, Box 8, f. 5; Dall to Carlile P. Patterson, Nov. 18, 1873, SIA, WHD, RU 7073, Box 4.

89. William Healey Dall to Benjamin Peirce, Feb. 22, 1874, Feb. 22, 1874, and Julius Hilgard to Dall, April 8, 1874, SIA, WHD, RU 7073, Box 4; Charles D. Sigsbee, *Deep Sea Sounding and Dredging: A Description and Discussion of the Methods and Appliances Used on Board the Coast and Geodetic Survey Steamer "Blake"* (Washington, D.C.: Government Printing Office, 1880); McConnell, *No Sea Too Deep*, 61, 68–70; Thomas G. Manning, *U.S. Coast Survey vs. Naval Hydrographic Office: A 19th-Century Rivalry in Science and Politics* (Tuscaloosa: University of Alabama Press, 1988), 38; Admiral W. Wharton, *Hydrographical Surveying* (London: J. Murray, 1898). The Lucas sounder remained in use for almost a century, serving as a bottom sampling device even after echo-sounders replaced line sounding for measuring depth.

90. Alexander Agassiz, *Three Cruises of the United States Coast Survey Steamer "Blake,"* vol. 1 (Boston: Houghton Mifflin, 1888), pp. vii–viii (the quote is on p. 37); Louis Agassiz to Spencer Baird, April 12, 1860, in Herber, ed., *Correspondence between Spencer Fullerton Baird and Louis Agassiz; Report of the Superintendent of the Coast Survey Showing the Progress of the Survey during the Year 1859*, 103; *Report of the Coast Survey Showing the Progress of the Survey during the Year 1879*, 3; George R. Agassiz, ed., *Letters and Recollections of Alexander Agassiz with a Sketch of His Life and Work* (Boston: Houghton Mifflin, 1913). Wire rope was not applied to dredging until after wire sounding devices were constructed, but it was developed in the 1830s for mining and produced commercially by the end of that decade. The Admiralty sanctioned its use for standing rigging. The clipper ships of the 1860s, whose enormous sails and high speeds produced heavier strains than hemp could withstand, required wire rigging. E. Rodney Forestier-Walker, *A History of the Wire Rope Industry of Great Britain* (London: The Federation of Wire Rope Manufacturers of Great Britain, 1952), 15, 23–24; Charles Singer, E. J. Holmyard, A. R. Hall, and Trevor I. Williams, ed., *A History of Technology* (Oxford: Clarendon Press, 1958), vol. 4, 592–593.

91. Dall to Peirce, Nov. 6, 1873, SIA, WHD, RU 7073, Box 4; Belknap to Dall, Aug. 5, 1874, SIA, WHD RU 7073, Box 8, f. 5. Jeffreys was quoted in Eric L. Mills, "One 'Different Kind of Gentleman': Alfred Merle Norman (1831–1918), Invertebrate Zoologist," *Zoological Journal of the Linnean Society* 68 (1980): 85.

92. F. S. L. Lyons, *Internationalism in Europe* (Leyden: A. W. Sythoff 1963); Helen M. Rozwadowski, "Internationalism, Environmental Necessity, and National Interest: Marine Science and Other Turn-of-the-Twentieth-Century Sciences," *Minerva* 42, no. 2 (2004): 127–149; Deacon, *Scientists and the Sea*, 367–373, 387.

6. Small World

1. Henrika Kuklick, *The Savage Within: The Social History of British Anthropology* (Cambridge: Cambridge University Press, 1991), 92. For the history of natural-history fieldwork, see David E. Allen, *The Naturalist in Britain: A Social History* (London: Allen Lane, 1976).

2. Billy G. Smith, *The "Lower Sort": Philadelphia's Laboring People, 1750–1800* (Ithaca: Cornell University Press, 1990); Marcus Rediker, *Between the Devil and the Deep Blue Sea: Merchant Seamen, Pirates, and the Anglo-American World, 1700–1750* (Cambridge: Cambridge University Press, 1987), 155–156; Anonymous, "American Deep-Sea Soundings," *Nature* 5 (1871–72): 324–325. Nina E. Lehrman notes the apprenticeship of orphan boys to merchant and other vessels during the nineteenth century in "From 'Useful Knowledge' to 'Habits of Industry': Gender, Race, and Class in Nineteenth-Century Technical Education" (Ph.D. diss., University of Pennsylvania, 1993).

3. Harold D. Langley, *Social Reform in the U.S. Navy* (Urbana: University of Illinois Press, 1967); Jonathan Neale, *The Cutlass and the Lash: Mutiny and Discipline in Nelson's Navy* (London: Pluto, 1985); Richard Henry Dana, Jr., *Two Years before the Mast: A Personal Narrative of Life at Sea* (New York: World Publishing Co., 1946).

4. Thomas H. Huxley to Henrietta Heathorn, Oct. 17, 1847, ICA, THH, Correspondence with Henrietta Heathorn, 2; George C. Wallich, "Lett's Diary for 1860," Sept. 13, 1860, NHM General Library, MSS WAL (hereafter cited as "Wallich Diary, 1860"); Herbert Swire, *The Voyage of the Challenger: A Personal Narrative of the Historic Circumnavigation of the Globe in the Years 1872–1876* (London: Golden Cockerel Press, 1938), vol. 1, 48–49; A. W. Brian Simpson, *Cannibalism and the Common Law: The Story of the Tragic Last Voyage of the "Mignonette" and the Strange Legal Proceedings to Which It Gave Rise* (Chicago: University of Chicago Press, 1984).

5. From Ralph Waldo Emerson, "English Traits" (1865), excerpted in Jonathan Raban, ed., *The Oxford Book of the Sea* (Oxford: Oxford University Press, 1993), 271–273.

6. Emily Dickinson, "Exultation is the Going," in R. W. Franklin, ed., *The Poems of Emily Dickinson, Variorum Edition* (Cambridge: The Belknap Press of Harvard University Press, 1998), vol. 1,183–184.

7. Basil Greenhill and Ann Gifford, *Travelling by Sea in the Nineteenth Century: Interior Design in Victorian Passenger Ships* (London: Adam & Charles Black, 1972); John Malcolm Brinnin, *The Sway of the Grand Saloon: A Social History of the Atlantic* (New York: Delacorte Press, 1971).

8. Henry N. Moseley, *Notes by a Naturalist: An Account of Observations made during the Voyage of H.M.S. "Challenger" round the World in the Years 1872–1876* (London: John Murray, 1892), 517; George R. Agassiz, ed., *Letters and Recollections of Alexander Agassiz* (Boston: Houghton Mifflin, 1913), 167.

9. George C. Wallich, "Daily Diary of Voyage of 'Bulldog,'" Oct. 10 and 11, 1860 ("sea was lashed" quote), Aug. 4, 1860 ("to a lands man" quote), NHM General Library MSS WAL (hereafter cited as "Wallich Journal"); Thomas H. Huxley to Henrietta Heathorn, May 12, 1849, ICA, THH, Correspondence with Henrietta Heathorn, 5.

10. William Healey Dall, Notebook No. 2, 1880, *Yukon*, Sept. 1, 1880, SIA, WHD, RU 7073, Box 24; Alfred M. Norman to Joshua Alder, June 1, 1863, and John Gwyn Jeffreys to Norman, June 25, 1868, NHM General Library, MSS ALD, vol. 5, 1054–1055, and vol. 4, 780.

11. Greg Dening, *Mr. Bligh's Bad Language: Passion, Power, and Theatre on the Bounty* (Cambridge: Cambridge University Press, 1992); Philip H. Rehbock, ed.,

At Sea with the Scientifics: The "Challenger" Letters of Joseph Matkin (Honolulu: University of Hawaii Press, 1992), 14.

12. The quotes are from Edward K. Calver to Hydrographer George H. Richards, Feb. 24, March 5, 9, 1869, HO, Surveyors Letters, or "S" Papers, 1869; Anita McConnell, *No Sea Too Deep: The History of Oceanographic Instruments* (Bristol: Adam Hilgar, 1982), 89.

13. Charles Wyville Thomson, *Voyage of the "Challenger": The Atlantic*, 2 vols. (London: Macmillan, 1877), vol. 1, 11.

14. Thomson, *Voyage,* vol. 1, 11.

15. B. J. Sullivan to Joseph Hooker [c. 1882], CUL Dar 107 f45r.

16. William Stimpson, "Journal of a Cruise in the US Ship Vincennes to the North Pacific Expedition, Behring Straits, Etc.," Nov. 22, 1853 ("comfortable" quote), May 11, 1854, Sept. 21, 1854 ("rolling" quote), Nov. 13, 1854, and Jan. 17, 1855, SIA, NPEE, RU 7253, Box 1 (hereafter cited as "Stimpson Journal"); Stimpson to Ringgold, April 7, 1853, SIA, NPEE, RU 7253, Box 1, folder 8.

17. "Wallich Journal," July 2, 3, 26, and Sept. 15, 1860 (quotes are from here).

18. Moseley, *Notes by a Naturalist*, 517.

19. Swire, *Voyage of the Challenger*, vol. 1, 10.

20. Chalres Erskine, *Twenty Years before the Mast* (Washington, D.C.: Smithsonian Institution Press, 1985), 211. Here is the mess bill for the wardroom from the U.S. Coast Survey Schooner *Yukon* in 1873, when the naturalist William Healey Dall led the surveying work (the forecastle menu is indicated parenthetically).

Mon: Beef, Peasoup, Canned Corn (same in Forecastle)

Tues: Pork, Beansoup, Beans, Macaroni (F = no macaroni)

Wed: Beef, Rice-soup, Tomatoes (F =Rice instead of Rice-soup)

Thurs: Pork, Green Peas, Duff (F = Potatoes instead of peas)

Fri: Beef, Bean-soup, Tomatoes (F = beans instead of tomatoes)

Sat: Pork, Pea-soup, Corn (F = same)

Sun: Beef or Oyster Soup, Beef, Green Peas, Duff (F = no soup; potatoes instead of peas)

"Schr. Yukon—1873—Mess Bill," SIA, WHD, RU 7073, Box 4.

21. The "Scurvy" quote is from "Wallich Journal," July 30, 1860; "Wallich Diary, 1860," Aug. 7 ("Salt Junk" quote is from here), and Aug. 15, 1860; Charles Wyville Thomson, "HMS Challenger: Diary of Sir Wyville Thomson," March 18, 1875, NHM Earth Sciences Library, Murray Col., Section 1, no. 10; hereafter cited as "Thomson Journal". On another occasion, Lieutenant Herbert Swire reflected on "the harrowed sensibilities of the last of the forty sheep they had taken aboard on that leg. Although he thought the animal deserved to live, having survived to be the last sheep facing slaughter, he admitted, "But we are selfish in what concerns our interior economy so I fear old fleecy-back must die"; Swire, *Voyage of the Challenger*, vol. 1, 167.

22. The quotes are from Moseley, *Notes by a Naturalist,* 501. See also Swire's descriptions of early dredge hauls in *Voyage of the Challenger*, vol. 1, 12, 15.

23. James H. Ferguson, "A Letter from Jack Staylight to His Old Shipmate 'Bill,'"

copied by Pelham Aldrich into his *Challenger* Journal on Feb. 17, 1913, RGS, LBR. MSS. AR 114A (hereafter cited as "Aldrich Journal"); Alexander Agassiz, *Three Cruises of the United States Coast Survey Steamer "Blake,"* vol. 1 (Boston: Houghton Mifflin, 1888), x; Rehbock, ed., *At Sea with the Scientifics,* 15.

24. The "drudging" quote is from George Campbell, *Log-Letters from the "Challenger"* (London: Macmillan, 1877), 482; Charles Wyville Thomson to Thomas H. Huxley, June 9, 1875, ICA, THH, vol. 27, 312; the "something good" quote is from Moseley, *Notes by a Naturalist,* 501; Captain George S. Nares to Hydrographer George Richards, Jan. 11, 1873, HO, Surveyor Letters, or "S" Papers, Miscellaneous Papers, 84a.

25. Edward K. Calver, "HMS 'Porcupine's' Atlantic and Mediterranean Journal, 1870," HO, Original Document Series, 395, pp. 13–15.

26. The philosopher" quote is from Henry Charnock, "HMS 'Challenger' and the Development of Marine Science," *The Journal of Navigation* 26, no. 1 (1973): 5 "Stimpson Journal," Feb. 24, May 24, June 10 (the "smell" quote is from here), Sept. 5, 1854, and April 23–26, 1855.

27. Charles Wright to Asa Gray, Nov. 4, 1853, Gray Herbarium, quoted in Ronald Vasile, "A Fine View from the Shore: The Life of William Stimpson," unpublished manuscript, chap. 2; William Stimpson to William Dall, Feb. 9, 1866, SIA, WHD, RU 7073; "Stimpson Journal," April 5, 1854.

28. The "paddle" quote is from Campbell, *Log-Letters,* 4–5; Thomson, *Voyage,* vol. 1, xviii–xix.

29. Thomson, *Voyage,* 114.

30. The quote is from "Wallich Journal," July 2, 1860; see also July 6 and 9, Aug. 19 and 20, 1860. Many of Dening's analyses of Bligh illuminate the problems Wallich had trying to accomplish scientific work on the *Bulldog* even though Wallich did not hold a position of power commensurate with Bligh's; Dening, *Mr. Bligh's Bad Language.*

31. The quotes are from "Wallich Journal," Sept. 7, 8, 9, 1860; "Wallich Diary, 1860," Sept. 8, 9, 13, 1860; Rehbock, ed., *At Sea with the Scientifics,* 53, 69; George S. Nares to George M. Richards, April 15, 1873, HO, Surveyor Letters, or "S" Papers, Miscellaneous Papers, 84a. Stealing and desertion were the most powerful forms of protest available to common sailors.

32. Wallich did preserve other samples this small. On one occasion, he trumpeted the value of carrying a naturalist on sounding voyages precisely because of his ability to tell if the sounder had really been at the bottom, based on the examination of such tiny amounts of sediment. "Wallich Journal," Sept. 2, 1860.

33. John Murray, "HMS 'Challenger' Diary of Sir John Murray," June 22, 1874, and Sept. 2, 1875 (the quote is from here), NHM Earth Sciences Library, Murray Col., Section 1, vol. 1 (hereafter cited as "Murray Journal"); Moseley, *Notes by a Naturalist,* 502; Rehbock, ed., *At Sea with the Scientifics,* 32, 75, 105.

34. "Murray Journal," July 15, 1875; "Stimpson Journal," Jan, 27, 1854.

35. The "experiment" quote is from Thomson, *Voyage,* vol. 1, xvii–xviii; see also x–xi, 3, 291. The "friendly" quote is from Eric Linklater, *The Voyage of the "Challenger"* (New York: Doubleday, 1972), 274. George S. Nares to George M. Richards, Jan. 26, 1874, HO, Surveyor Letters, or "S" Papers, Miscellaneous Papers, 84a.

36. Spencer F. Baird to Cadwallader Ringgold, Nov. 12, 1852, SIA, NPEE, RU 7253, Box 1, folder 2; "Wallich Journal," July 2, 1860; Charles Bright to Cyrus W. Field [May 1858], C&WA, Atlantic Telegraph Company Collection, Box D, Atlantic Telegraph Company Letterbook, reference 10066, #437; Edward K. Calver, "HMS 'Porcupine' Atlantic Journal, 1869," HO, Original Document Series, 394, 49.

37. "Stimpson Journal," Oct. 19, 1854. For examples of scientists' descriptions of shipboard dinners, see John James Wild, "HMS 'Challenger': Diary of Mr. J. J. Wild," vol. 1, Dec. 25, 1872, NHM Earth Sciences Library, Murray Col., Acc. no. 179712, Section 1, no. 6–9 (hereafter cited as "Wild Journal"); "Thomson Journal," Dec. 25 and 31, 1873; and Swire, *Voyage of the Challenger*, vol. 1, 125–127. Brinnin, *Sway of the Grand Saloon*, 179–180; John F. Kasson, *Rudeness and Civility: Manners in Nineteenth-Century Urban America* (New York: Hill and Wang, 1990), 182–214.

38. "Wild Journal," Dec. 25, 1872; "Thomson Journal," Dec. 25, 1873 (long quote); Swire, *Voyage of the Challenger*, vol. 1, 125–127. The toast is from "Thomson Journal," Dec. 31, 1873.

39. "Stimpson Journal," May 3, 1854; Sidney W. Mintz, *Sweetness and Power: The Place of Sugar in Modern History* (New York: Penguin Books, 1985).

40. "Murray Journal," July 17, 1873; the "shook" quote is from "Aldrich Journal," Sept. 18, 1873; Swire, *Voyage of the Challenger*, vol. 1, 119.

41. Swire, *Voyage of the Challenger*, vol. 1, 13, and vol. 2, 143; Horace Beck, *Folklore of the Sea* (Middletown, CT: Wesleyan University Press, 1973). These sea chanteys were not written down until the second half of the nineteenth century when observers began collecting the folklore of sailing ships. This impulse appeared on the heels of the growth of anthropological efforts to preserve disappearing native cultures and civilizations.

42. Swire, *Voyage of the Challenger*, vol. 2, 161; "Aldrich Journal," Dec. 31, 1872.

43. Swire, *Voyage of the Challenger*, vol. 1, 142; "Murray Journal," April 1, 1875; Rudolph von Willemoes-Suhm to Thomas H. Huxley, Sept. 2, 1874, ICA, THH, vol. 29. The *Challenger* carried sixteen breech-loading guns and rifles as well as an air gun for sporting purposes. See *List of Scientific Instruments, Dredging and Sounding Apparatus, Chemicals, Etc. on Board HMS "Challenger"* (London: HMSO, 1873), held at HO, Hydrographic Office Publication 43.

44. "Murray Journal," March 13, 1875; Andrew F. Balfour, HMS *Challenger* Journal, Aug. 30, 1874, RGS, Library Mss. 113 B, vol. 1; Swire, *Voyage of the Challenger*, vol. 1, 59; "Aldrich Journal," Aug. 27, 1873.

45. Swire, *Voyage of the Challenger*, vol. 1, 61–63; "Aldrich Journal," Sept. 4, 1873; "Thomson Journal," Dec. 31, 1873; "Murray Journal," Aug. 30, 1874; Margaret S. Creighton, *Rites and Passages: The Experience of American Whaling, 1830–1870* (Cambridge: Cambridge University Prss, 1995), 117–123.

46. "Wallich Journal," Sept. 9, 1860; Beck, *Folklore of the Sea;* Patricia Ann Carlson, ed., *Literature and Lore of the Sea* (Amsterdam: Rodolphi, 1986).

47. "Stimpson Journal," Aug. 25 (quote), and Nov. 9, 1853; Frederick J. Stuart, Sr., NPEE Journal, Aug. 28, 1853, NA, Microscopy 88, roll 7, 45. Swire also noted that the old sailors would ascribe foul weather to the killing of storm petrels (also known as Mother Carey's chickens), in *Voyage of the Challenger*, vol. 1, 38. "Murray Journal," July 25, 1875.

48. Thomas H. Huxley to Henrietta Heathorn, HMS Bulldog, May 7, 14, 1848, ICA, THH, Correspondence with Henrietta Heathorn, 19.

49. The "case" quote is from "Notes," *Nature* 7 (1872–73): 131; Swire, *Voyage of the Challenger*, vol. 1, 14; Charles Wyville Thomson to Henry H. Huxley, May 13 [1873], ICA, THH, vol. 27, 301; Vasile, "A Fine View from the Shore," chap. 2; George Wallich to Joseph Hooker, June 1, 1860, KEW, English Letters (1857–1900), vol. 104, 348; Thomson, *Voyage*, vol. 1, 18; "List of Books for HMS 'Challenger,'" Appendix E in Rehbock, ed., *At Sea with the Scientifics*, 359–370.

50. Thomas H. Huxley to Henrietta Heathorn, Oct 14, 1849, ICA, THH, Correspondence with Henrietta Heathorn, 72; William Healey Dall to John G. Jeffreys, Oct. 6, 1872, RSA, MC 9.414; Rehbock, ed., *At Sea with the Scientifics*, 17–18.

51. Swire notes, with evident pride, that the indian-ink sketches he had made for the charts were to be sent home for engraving, in Swire, *Voyage of the Challenger*, vol. 1, 168. The importance of painting on English voyages of exploration dates back to the late sixteenth century; see Susan Schmidt Horning, "The Power of Image: Promotional Literature and Its Role in the Settlement of Early Carolina," *North Carolina Historical Review* 70, no. 4 (1993): 373–374. Anne L. Larsen, "Not Since Noah: The English Scientific Zoologists and the Craft of Collecting, 1800–1840" (Ph.D. diss., Princeton University, 1993), 209, 216–219.

52. Hart Gimlette, "Tables of Temperature, Density of Sea Water, Soundings and Observations on Specimens Taken at Various Depths in the Atlantic on Board the 'Cyclops' during June and July 1857," 21, in Thomas H. Huxley, "Deep Sea Soundings," Notebook 116, p. 20, ICA, THH, Scientific Notebooks, 2nd ser., 116; "Wallich Journal," Aug. 6, 1860; Swire, *Voyage of the Challenger*, vol. 1, 158–160, 168; "Aldrich Journal."

53. *Challenger Sketchbook: B. Shephard's Sketchbook of the HMS Challenger Expedition, 1872–1874*, ed. Harris B. Stewart, Jr., and J. Welles Henderson (Philadelphia: Philadelphia Maritime Museum, 1972); Beck, *Folklore of the Sea*, 187. Compared to other mariners, whalers were especially productive shipboard artists because their periods of hard work alternated with long periods of idleness.

54. "Murray Journal," April 9, 1874, and May 12–30, 1874; "Aldrich Journal," Sept. 18, 1873.

55. Huxley to Heathorn, May 7, 14, 1848;. Henry N. Moseley to Captain Evans, Sept. 6, 1874, HO, Surveyor Letters, "S" Papers, Miscellaneous Papers, 84b; William Healey Dall to Carlile P. Patterson, Nov. 3, 1872, SIA, WHD, RU 7073, Box 4.

56. "Wallich Journal," Nov. 9, 1860; Sylvanus Bailey to William Healey Dall, April 22, 1875, July 9, Oct. 1, 1876, SIA, WHD, RU 7073, Box 7, folder 1; Robert M. King to Thomas H. Huxley, July 8, 1863, ICA, THH, vol. 19, 155; Herbert Swire to G.B.C, in "Memoir to Henry N. Moseley," in Moseley, *Notes by a Naturalist*, x–xi.

57. The "somehow" quote is from "Aldrich Journal," undated entry after final *Challenger* entry, Dec. 10, 1875; the "head" quote is from Huxley, "Deep Sea Soundings," Notebook 116, p. 1; the "jellyfish" quote is from Huxley, "Mss. Voyage of the HMS Rattlesnake Notebook," No. 2 (1847–48), Dec. 20, 1847, ICA, THH (the emphasis is mine); the "persisted" quote is from Swire, *Voyage of the Chal-*

lenger, vol. 1, 17–18; Thomson to Huxley, May 13 [1873]. Larsen discusses the commodification of scientific specimens in "Not Since Noah," 288–355, as does Debra Lindsay in *Science in the Subarctic: Trappers, Traders, and the Smithsonian* (Washington, D.C.: Smithsonian Institution Press, 1993), 89–104. William R. Stanton, *The Great United States Exploring Expedition of 1838–1842* (Berkeley: University of California Press, 1975), 262.

Epilogue

1. The quote is from John R. Stilgoe, *Alongshore* (New Haven: Yale University Press, 1994), 30. Recently, scholars have begun to see the beach—the zone including the edge of the land and the beginning of the sea—as a place whose history is worth telling. Greg Dening, *Islands and Beaches: Discourse on a Silent Land: Marquesas, 1774–1880* (Honolulu: University of Hawaii Press, 1980); Alain Corbin, *The Lure of the Sea: The Discovery of the Seaside in the Western World, 1750–1840,* trans. Jocelyn Phelps (Cambridge: Polity Press, 1994); Lena Lenček and Gideon Bosker, *The Beach: The History of Paradise on Earth* (New York: Viking, 1998). The 1995 Smithsonian Institution exhibit Ocean Planet presents results of a study of the health of the world's oceans, raising awareness of the mutual dependency between humans and the ocean, but failing to consider the historical dimension of that relationship; see http://seawifs.gsfc.nasa.gov//ocean_planet.html.
2. Clark G. Reynolds, "British Strategic Inheritance in American Naval Policy," in Benjamin W. Labaree, ed., *The Atlantic World of Robert G. Albion* (Middletown, CT: Wesleyan University Press, 1975), 169–194; Isabel Ollivier, "Pierre-Adolphe Lesson, Surgeon-Naturalist: A Misfit in a Successful System," in Roy MacLeod and Philip F. Rehbock, ed., *Nature in Its Greatest Extent: Western Science in the Pacific* (Honolulu: University of Hawaii Press, 1988), 45–64.
3. The *Daily Telegraph,* May 2, 1857, in IEE, Press Cuttings Relating to Telegraphy, UK0108 SC MSS 021, "Telegraphic Notes, 1839–1857," vol. 1.
4. John Murray and Johan Hjort, *The Ocean: A General Account of the Science of the Sea* (New York: Henry Holt, 1913); Charles Wyville Thomson, "The Cruise of the 'Knight Errant,'" *Nature* 22 (1880): 405–407.
5. Margaret Deacon, *Scientists and the Sea, 1650–1900: A Study of Marine Science,* 2nd ed. (Brookfield, VT: Ashgate Publishing Co., 1997), x–xiv.
6. Daniel R. Headrick, *The Invisible Weapon: Telecommunications and International Politics, 1851–1945* (Oxford: Oxford University Press, 1991).
7. John G. Jeffreys and William B. Carpenter, "The 'Valorous' Expedition: Papers by Dr. Gwyn Jeffreys and Dr. Carpenter, CB, FRS," *Proceedings of the Royal Society* 25, no. 173 (1876): 177–237; Roger R. Revelle, "Director of Scripps Institution of Oceanography, 1951–1964," an oral history conducted in 1984 by Sarah Sharp, Regional Oral History Office, The Bancroft Library, University of California, Berkeley, 1988, Scripps Institution of Oceanography, Reference Series no. 88–20, Nov. 1988, pp. 19–20. For a review of postwar ocean science in the United States, see Ocean Studies Board, National Research Council, *50 Years of Ocean Discovery: National Science Foundation, 1950–2000* (Washington, D.C.: National Academy Press, 2000).

8. The "always" quote is in Sonya Senkowsky, "Charting New Territory: Alaska's Seamounts," *BioScience* 52, no. 11 (2002): 968–976. David Malakoff and Charles Seife, "Into the Unknown," *Science* 290 (Oct. 6, 2000): 25; David Malakoff, "Marine Researchers Hope to Sail Off into the Unknown," *Science* 296 (May 24, 2002): 1386–1387.

9. With satellite technology, this firm link between oceanography and seagoing has recently begun to change.

10. Chandra Mukerji amply makes this point in *A Fragile Power* (Princeton: Princeton University Press, 1989).

Acknowledgments

THE TROPE OF VOYAGING is obvious yet appropriate for the journey that resulted in this book. I am pleased to have the opportunity to acknowledge some of the many people who helped me with this project. Its faults are my own, despite my many debts.

This book had its genesis on a research voyage I made as an undergraduate, to the North Atlantic aboard the RV *Westward,* a school ship of the Sea Education Association (SEA) Semester program, in Woods Hole, Massachusetts. From my shipmates and teachers I learned about oceangoing life and work, especially the inherent contradictions that arise, even today, from trying to do scientific research at sea.

My historical research took me to many places on the two continents flanking the Atlantic. In the United States I found helpful assistance from many librarians and archivists at the the American Philosphical Society, Philadelphia, and in Washington, D.C., at the Library of Congress, the National Archives, the National Observatory, and throughout the Smithsonian Institution, including its Institution Archives and Library, the Dibner Rare Book Library, and the National Museum of American History. An extended stay in Britain gave me the opportunity to conduct research at several institutions whose staff likewise provided tremendously valuable aid: the Cable and Wireless Archives, London; the Cambridge University Library, Manuscripts Department; the Edinburgh University Library, Special Collections Department; the Hydrographic Office, Taunton; the Imperial College Archives, London; the Institution of Electrical Engineers, London; the Natural History Museum, London; the Royal Botanic Gardens Archives, Kew; the the Royal Geographical Society, London; the Royal Society, London; the Science Museum, London; the Ulster Museum, Belfast; the University College Library, London; and the Wellcome Institute for the History of Medicine, London. For assistance and friendship far beyond the call of duty, I am especially grateful to Anne Barrett, Imperial College Archives, and the late John Thackray, the Natural History Museum Archives.

Several people provided me with invaluable resources, references, images, or other critical assistance. I heartily thank Raymond Manning, William Deiss, and Ronald Vasile for access to North Pacific Exploring Expedition journals, including their transcription of William Stimpson's; Richard MacAndrew, for kindly giving me a copy of his privately published biography of Robert MacAndrew; and Peter S. Davis for his help with information about northern English naturalist-dredgers and the Northumberland perspective on the *Porcupine* voyage. Thanks also to Molly Berger, Matthew Burnside, Steven Dick, Joy Harvey, Sue Helper, John Krige, Anita McConnell, Gregory Nobles, Jim Secord, and Gary Weir, for generous advice, suggestions, and assistance that added important dimensions to this book. Other colleagues have commented on papers, spent time informally discussing my project, and generally provided inspiration, including Keith Benson, James Fleming, Mats Fridlund, Elizabeth Greene-Musselman, Myles Jackson, Jordan Kellman, Gary Kroll, Naomi Oreskes, Michael Osborn, Philip J. Pauly, the late Phillip F. (Fritz) Rehbock, and Ronald Rainger.

In my travels I have had the good fortune to find a number of institutions where my work was nurtured, challenged, and enhanced. My teachers in the Department of History and Sociology of Science at the University of Pennsylvania encouraged me to cast a wide net as I contemplated a topic that covered a longer swath of time than many historical works. I thank, in particular, Robert E. Kohler and Henrika Kuklick for their patient guidance and their confidence. Svante Linqvist gave me important advice during a semester he spent at the department. I also want to thank my student colleagues, not all Penn students, who helped define the boundaries of my topic: Christopher Feudtner, Lissa Hunt, Anne Kleffner, Nina Lehrman, Stuart McCook, Elizabeth Toon, Jennifer Tucker, and Rebecca Ullrich.

I also had the good fortune to spend time at the Smithsonian Institution, where I similarly profited from experienced scholars as well as "fellow fellows," especially Laura Edwards, John Hartigan, Laura Helper, Marc Rothenberg, Peggy Shaffer, and Walter van de Leur. Pamela Henson deserves special recognition for her unstinting willingness to help and the depth of her knowledge. This book owes much to her aid. I also benefited from time I was lucky enough to spend with the Department of History and Philosophy of Science at Cambridge University, the Centre for the History of Science, Technology, and Medicine at Imperial College, London, the History Department at Case Western Reserve University, and, last though hardly least, the Tuesday night Irish Times group and the London-based Sarah Hughes Newting Society. Recently I had the pleasure of joining the History Department at the University of Connecticut, where I found a new group of congenial and stimulating colleagues as I finished this project.

Many people in addition to those named above, and no doubt including some I have forgotten to name, helped me tackle the challenge of telling this story. Their criticism sharpened my thinking and their encouragement kept me working. I must start with Eric Mills, who is the guardian spirit of the history of oceanography, nurturing the field and encouraging other historians to enter it. He has been unfailingly generous with his time, his considerable knowledge, and his constant encouragement. He read the whole manuscript and made innumerable improvements to it. I am also grateful to Margaret Deacon and D. Graham Burnett for generous, rigorous, and insightful comments on the manuscript. At Harvard University Press, Ann Downer-

Hazell offered brisk encouragement from start to finish, while Nancy Clemente smoothed rough edges and gently reminded me that not everyone knows what a tryworks is. Others who read all or parts of the manuscript offered me the benefit of their wisdom and experience, notably Betsy Hanson, Anne Larsen-Hollerbach, and Shari Rudavsky (who also contributed enthusiastically to the list of rejected titles). It is too late to thank personally David van Keuren, my friend and long-time colleague, now deceased, who is missed by all who knew him, including me.

Research for this project was supported by a National Science Foundation Dissertation Improvement Grant (Ref: SBR93–20247), a Dibner Library Research Scholarship, and a Smithsonian Institution Fellowship. A National Endowment for the Humanities Dissertation Grant (Ref: FD–22185–94) funded a year of writing and the William E. & Mary B. Ritter Fellowship from the Scripps Institution of Oceanography came at an important juncture. And earlier version of Chapter 6 appeared as "Small World: Forging a Scientific Maritime Culture for Oceanography," *Isis* 87 (1996): 409–429, published by the University of Chicago Press; © by the History of Science Society.

Personal debts are the most satisfying of all to discharge. Many of those mentioned above are also friends, and some friends singled out here are also colleagues. Foremost among these is Jennifer Gunn, both friend and colleague, in the best senses of these words, since our first days of graduate school. Pete Daniel, Jackie Hutter, Matt Payne, and Dana Randall and have lent me support and encouragement at critical times. Reaching farther back in time, and also continuing strongly in the present, my family provides the home port from whence all my journeys depart and return. My siblings, Jeanne, John, and Annie, are a constant source of strength and balance. We grew up on Lake Erie, the inland sea that taught me to love the ocean. My children, Thad and Meg, who are growing up by salt water, help me see the world anew every day. It is my parents, though, to whom I dedicate this book, for their boundless love.

Index

Barlee, George, 125, 126, 135, 147
Barnett, Edward, 75
Barnum, P. T., 11
Barron, Samuel, 33
Bathybius haekelii, 163–164, 166
Bathymetrical charts, 78, 79–80, 81, 85, 87, 95
Beachcombing, 34, 52, 99, 100, 104, 107, 109, 144, 216
Beaches, 3–4, 7–9, 107, 179, 217
Beacon, 57, 107, 112, 135, 138, 231n32
Beagle, 48, 186, 187, 215, 235n12
Beaufort, Francis, 48, 49, 58, 65, 94–95
Belknap, George E., 34, 170, 172
Berryman, Otway A., 81–82, 83–84, 85, 86, 91
Bibb, 167
Biological science. *See* Marine zoology
Blake, 172, 190
Botany, 32, 129, 148, 165
Bottom profiles, 79–80, 89–90, 93
Bottom samples: and whale distribution, 43; from the deep ocean, 50, 54, 190; and navigational surveying, 53, 70; microscopic study of, 63, 140, 144, 163, 205; and submarine telegraphy, 84, 86, 88–92; as confirming the sounding, 85, 93
Bounty, 183
Bowditch, Nathaniel, 71
Brady, George S., 106, 114
Brady, Henry B., 114
Brassey, Anne, 25, 120, 121, 128
Bright, Charles, 24
Brighton, 8, 9
Britain, 6, 47–48, 50, 58, 92, 99, 130, 214, 215
Britannia, 10
British Association for the Advancement of Science: U.S. contact with, 56; dredging as a national project, 65, 99, 109; dredging activities of, 111–117, 128, 143, 145, 146–148, 167, 226n64; use of yachts, 121–126; meetings, 144, 156; use of steamers, 149 153; and *Challenger* plans, 160
British Association for the Advancement of Science dredging committees. *See* Dredging committees
British Museum, 112, 138, 139, 242n30
Brooke, John M., 50, 51, 53–55, 62, 84–85
Brooke's sounder, 50, 54, 56, 69, 84–86, 87, 151
Buchanan, John Y., 166, 198
Bulldog, 26 45, 100, 165; and submarine telegraphy, 58, 93, 95; planning for the voyages, 59, 196, 204; Wallich's journal, 60,

181; activities during voyages, 93, 140–141, 182, 186, 189, 193–194, 205
Byron, George Gordon, Lord, 211, 213

Cable laying: by Great Eastern, 12; role in drawing attention to the ocean, 13, 14, 16, 24, 89, 215, 238n50; 1857 and 1858 attempts, 92, 93; and scientists, 165, 196
Calver, Edward, 151, 152, 156–158, 168, 183, 191, 196
Campbell, George C., 164, 200
Cape Horn, 22, 182
Captains: and the difficulties of deep-sea research, 77; in the yachting industry, 120; understanding of scientists' objectives, 156; and crew, 178; role and place on ships, 184, 186, 189, 190, 195
Carpenter, William B.: and argument for government patronage, 30, 145–146, 148, 150, 153–154, 168; interest in voyaging, 61–62; and dredging, 109; plan for a government dredging survey, 142, 144–147, 148; and *Porcupine* work, 155–156, 159–160, 196; and oceanic circulation, 155, 160, 161, 162, 163, 251n51; and the *Challenger* expedition, 161–163, 165–166, 167
Challenger expedition, 27–29, 32, 45, 48, 61, 140, 164, 169–170, 179, 180, 188, 190; planning of, 30, 136, 161–166, 167; *Challenger* Office and *Reports*, 114, 173, 216; and oceanography, 168; and national pride, 173; legacy of, 177; use of space on the ship, 183–184, 185, 187; shipboard life and work, 189, 197–201; the ship as scientific workplace, 193, 194–196, 203–208
Charting, 53, 62, 64, 70, 71–73, 215. *See also* Bathymetrical charts
Chemistry, 114, 155, 159, 162, 166, 173, 218
Chicago Academy of Sciences, 56
Children, 8, 17, 104, 106, 109, 123, 128, 246n78
Christmas, 188, 197–198, 199
Circulation, oceanic, 121, 136, 155, 160, 161, 162, 163, 251n51
Civil War, 14, 44, 55, 56, 92, 94, 167
Coleridge, Samual Taylor, 202
Collections: importance of to naturalists, 57, 110, 111, 124; as community resources in natural history, 138, 143, 242n30, 249n21
Common seamen: writing by, 4, 19, 205; reputation of, 9, 178–179; interactions with scientists, 32–33, 177, 187, 192, 193–194, 201, 204; and strenuousness of ocean research, 77, 127, 135, 149, 153, 190, 191; in-

volvement in scientific work, 184, 191, 200, 203, 208; shipboard life and work, 186, 188, 190, 195, 197, 198, 199, 206; and stealing as protest, 193–194, 195, 258n31

Congress, 78

Conrad, Joseph, 22, 217

Continental shelf, 71

Cook, James, Captain, 20, 39, 46, 47, 50

Cooper, James Fenimore, 21

Cordelia, 150

Correspondence networks, 111, 128, 143, 242n29

Cowes, 17, 118

Crimean War, 94

Crinoids, 144–145, 147, 148

Crusoe, Robinson, 7, 28

Cyclops, 95

Dall, William Healey: consultations with other naturalists, 56, 116, 192, 204; efforts to combine hydrography with marine zoology, 58, 143–144, 170, 207; and sounding, 172, 173, 182, 208

Dalrymple, Alexander, 65

Dana, James D., 51

Dana, Richard Henry: writing and voyage experience, 21, 107, 217; and the creation of a new image of seagoing, 22, 23, 62, 182, 213; and reform, 178

Darwin, Charles: and voyage narratives, 20, 28, 29; on voyaging and science, 48, 58, 59, 205; and natural history, 100, 110, 115; evolutionary theory of, 147, 163; on the *Beagle,* 186, 187, 215, 235n12

Davis, J. E., 150, 151

Dayman, Joseph, 95, 151, 164, 179, 205

Deacon, Margaret, 217

Deep-sea dredging: Jeffreys as pioneer of, 35, 126–127, 143, 156–158; hydrographers' contribution of technology for, 136, 149–153, 154–155, 157; and the question of life at depths as an incentive for, 136–142, 158; difficulty of, 159, 191

Deep-sea sounding: early efforts, 6, 48–49, 50, 51; debates over possibility of, 32; importance of for zoology, 54; soundings as experiments, 54, 75–79, 151; as a national accomplishment, 64, 94; first organized program of, 69, 70, 74–82, 83–84, 85–86; time and effort expended in, 150, 153; application of steam power to, 150, 153, 155, 156, 158, 194

Defoe, Daniel, 28

de la Beche, Henry, 57

Denham, Henry, 33

Depot of Charts and Instruments (U.S.), 44, 71–72

Desertion, 77, 193–194, 195, 206, 258n31

Dickens, Charles, 11

Dickinson, Emily, 179–180

Discipline on ships, 13, 77, 177, 178–179

Diving, 33, 193

Dogger Bank, 34, 125

Dolphin, 4, 85; goals of cruises of, 24; report of, 73, 94; activities during cruises, 77, 78, 81; and submarine telegraphy, 82, 91

Drawing and painting at sea, 19, 197, 205–207, 208

Dredging: use of yachts, 25, 117–118, 119, 121–126, 128, 135–136; depths dredged, 34; in the North Pacific Exploring Expedition, 50–53; from a ship versus a small boat, 52; sociability of, 97, 106–107, 122, 128, 241n19; popularity of in northern waters, 99; earliest dredging efforts, 100–103; technique of, 103, 117; communities of practitioners, 111–117, 136; blank forms for, 116, 123, 243n41; appeal for naval assistance for, 149–150

Dredging committees: role in bringing naturalists together, 56, 112–113, 146; and promotion of ocean science, 109, 145, 148; and vessels for dredging, 122–126, 149, 226n64

Dredging technique: sharing of with others, 56–57, 126–127; and Jeffreys, 143, 156–158; on *Porcupine,* 157–158

Dredging technology: dredges as sampling devices, 33; industrial origins of, 34, 100, 101–102, 126; diffusion of, 56–57, 143; dredges as naturalists' preferred tool, 101, 102, 137, 140; hempen tangles, 158, 168–169; and *Challenger* as technologically conservative, 168; wire rope, 168, 171, 172, 255n90; deep-sea trawling, 169–170; loss of gear, 194–195. *See also* Deep-sea dredging

Dutch seascape, 7, 20

Ehrenberg, Christian G., 63, 86, 101, 140, 141

Emerson, Ralph Waldo, 179

Emigration, 3, 4, 6, 9–11, 13, 178

Engineers: telegraph, 23, 82, 89, 196, 215; and use of the ocean, 70, 218; on ships, 153, 183, 190, 191, 194

Environmental crisis of the oceans, 213

Erskine, Charles, 23, 208

Evolutionary theory, 110, 147, 163–164

marine zoology, 108, 142, 145–146; and the *Challenger* expedition, 162, 163, 164–165, 166; and the *Rattlesnake* voyage, 179, 182, 207, 208, 215

Hydra, 151–153

Hydrographers: role of in making the ocean an object of study, 15, 33, 55–56; adoption of voyage narratives, 23; conception of sea-floor shape, 70, 74; and skill, 75, 78, 80–82

Hydrographic Office (British): adoption of Brooke's sounder, 56, 151; responsibility for survey and exploration voyages, 58; background on, 64–65; and deep-sea research, 94–95, 136, 144, 166, 192; and fostering of technological developments, 150, 152–153

Hydrography: charting, 6, 214, 233n3; surveys for prospective cables, 14–15, 23, 58, 92–94, 153, 170; contribution to ocean science, 19, 214; contribution to deep dredging, 35, 149–153, 168, 170; and Brooke, 51, 53–54, 56; pioneering of deep sounding, 55, 57, 76, 94, 95, 167, 214; as a national achievement, 62, 65, 173; practices for charting, 71–73, 205; and bottom sampling, 84, 163; and life at great depths debate, 136–138; and work at sea with scientists, 177, 184, 186, 190, 195, 207, 208

Hyndman, George C., 114, 122

Icebergs, 91, 205–206

Immigration. *See* Emigration

Imperialism, 40, 213, 214–215

Indian Ocean, 40

Industrialization, 213

Internationalism, 173

Irving, Washington, 21

Isle of Wight, 118

Jameson, Robert, 100

Japan, 51

Jeffreys, John Gwyn: choice of voyage narrative format, 25; comparison of ocean study to astronomy, 30; as pioneer of deep dredging, 35, 126–127, 143; and dredging, 109, 114, 124–127, 148, 183; and marine zoology, 110, 135, 249n21; and women in natural history, 128–129; role in promoting ocean science, 142, 143–144, 204–205, 217; and *Porcupine* work, 155–158, 160, 196; and the *Challenger* expedition, 162, 163, 165; and government patronage, 167; and national pride, 173

John Adams, 33, 75

Johnston, George, 100, 106, 112, 113, 114, 128

Journalism: role in drawing new attention to the ocean, 11–12, 13, 17, 119; and submarine telegraphy, 15, 24, 88, 89–90, 92, 215, 216

Juan Fernandez, 28

Kennedy, John Pendleton, 51

Kingsley, Charles, 104

Knight, Edward F., 25

Lamarck, Jean-Baptiste de Monet de, 100

Landlubbers: reasons for going to sea, 5–6, 20; interest in the deep ocean, 9; experience of seagoing, 180, 182, 188–189, 200–201, 208, 216

Laughrin, William, 115, 127, 156

Lebour, Marie V., 130

Lee, Samuel P., 24, 53, 77–78, 81, 84, 94

Lewis and Clark expedition, 47, 58

Life at great depths debate, 26, 91–92, 136–142, 158, 162, 190. *See also* Azoic theory

Life at sea, for passengers, 12–13

Lightning, 167, 251n46; and deep dredging, 35, 61, 154–155, 157; and government patronage, 131, 136, 161; and promotion of ocean science, 159, 163, 166, 168

Lind, Jenny, 11

Linnean Society, 115, 124, 127, 139, 143, 156, 160

Linneaus, 99–100

Literacy, 18, 19, 23

Logbooks, 18–19, 26, 27, 121, 228n11

Lyell, Charles, 112, 139, 143, 162, 163

MacAndrew, Robert, 25, 122–124, 125, 127

Magnetism, 5, 31, 40, 62, 71

Maine, 172

Malacology, 99–100, 106, 111, 112, 124, 135, 143, 144

Marine Biological Laboratory (U.S.), 130–131

Marine natural history: role in making the ocean an object of study, 33, 55–56; collections from shallow water, 34; beachcombing, 99; middle-class pursuit of, 103–108; and yachting, 121; popularity of, 215, 217

Marine zoology: yachtsmen and, 25; North Pacific Exploring Expedition achievements in, 57; as a new field, 60, 110–117, 149, 177; and national pride, 64, 173; sustained program in Britain, 99, 109, 123; dredges as